VAKE

백승도 지음

오늘도 솔드아웃!

VEGAN BAKING

베이크의 비건 베이킹

길벗

VAKE VEGAN BAKING

베이크의 비건 베이킹

초판 발행 · 2022년 5월 3일

지은이 · 백승도
발행인 · 이종원
발행처 · (주)도서출판 길벗
출판사 등록일 · 1990년 12월 24일
주소 · 서울시 마포구 월드컵로 10길 56(서교동)
대표전화 · 02) 332-0931 | **팩스** · 02)323-0586
홈페이지 · www.gilbut.co.kr | **이메일** · gilbut@gilbut.co.kr

편집팀장 · 민보람 | **책임편집** · 서랑례(rangrye@gilbut.co.kr) | **제작** · 이준호, 손일순, 이진혁
영업마케팅 · 한준희 | **웹마케팅** · 김선영, 류효정 | **영업관리** · 김명자 | **독자지원** · 윤정아

디자인 · 박찬진 | **영문 번역** · 박근일(Ken park), 신소영 | **교정교열** · 이정현
푸드스타일링 · 정재은 | **사진** · 장봉영 | **사진 어시스턴트** · 박효정, 신지우
CTP 출력 · 인쇄 · 제본 · 상지사 피앤비

ISBN 979-11-6521-959-8(13590)
(길벗 도서번호 020199)

정가 20,000원

독자의 1초까지 아껴주는 정성 길벗출판사
(주)도서출판 길벗 | IT실용, IT/일반 수험서, 경제경영, 취미실용, 인문교양(더퀘스트) www.gilbut.co.kr
길벗이지톡 | 어학단행본, 어학수험서 www.eztok.co.kr
길벗스쿨 | 국어학습, 수학학습, 어린이교양, 주니어 어학학습, 교과서 www.gilbutschool.co.kr
페이스북 · www.facebook.com/gilbutzigy | **트위터** · www.twitter.com/gilbutzigy

"

독자의 1초를 아껴주는 정성!
세상이 아무리 바쁘게 돌아가더라도
책까지 아무렇게나 빨리 만들 수는 없습니다.
인스턴트식품 같은 책보다는
오래 익힌 술이나 장맛이 밴 책을 만들고 싶습니다.

땀 흘리며 일하는 당신을 위해
한 권 한 권 마음을 다해 만들겠습니다.
마지막 페이지에서 만날 새로운 당신을 위해
더 나은 길을 준비하겠습니다.

독자의 1초를 아껴주는 정성을 만나보십시오.

"

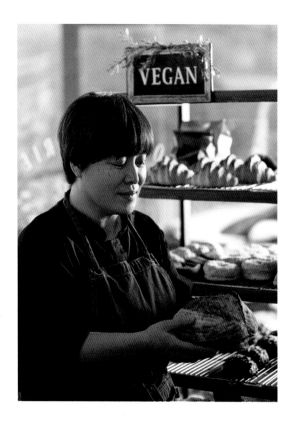

저는 10년 넘게 키스 더 케이크라는 공방을 운영하며 버터크림 플라워 케이크를 만들고 가르치는 일을 했어요. 짤주머니에 색색의 버터크림을 넣어서 움직이다 보면 내 손에서 예쁜 크림 꽃이 피어나고, 그 꽃으로 원하는 디자인의 케이크를 만드는 설레는 일이어서 할머니가 되어도 케이크를 만들며 살아야겠다고 생각했죠.

그러다 몇 년 전 프랑스 여행에서 단순한 호기심으로 단기 빵 수업을 들었어요. 시설이나 장비를 제대로 갖추지 않은 수업이었는데, 프랑스 명장님의 빵에 대한 열정과 오븐에서 마술같이 순식간에 부풀어 오르는 바게트를 보고 한순간에 매료되어 빵을 제대로 배우고 싶다는 생각이 들었고, 한국으로 돌아와 INBP에서 프랑스 빵을 배웠어요.

전 이 과정만 끝내면 빵에 대해 완벽히 이해하고 잘할 수 있을 줄 알았는데, 빵은 같은 레시피여도 작업 환경에 따라 결과물이 달라지더라고요. 종종 결과물이 마음에 들지 않을 때도 있고 뭐가 문제인지 모를 때도 있는데, 너무 아이러니하게 이렇게 잡힐 듯 잡히지 않는 빵 작업이 저에게는 지루하지 않게 인생을 보내고 있는 것처럼 느끼게 해줘서 빵을 만들면서 살 제 인생이 점점 더 기대가 돼요.

제 비건빵은 유제품 섭취에

어려움을 느끼는 분이나

동물복지를 실천하시는 분 등

엄격한 채식주의자도 먹을 수 있고,

논비건이 먹어도 맛있는

빵을 만드는 것을 목표로 삼고 있어요.

더 많은 연습과 노력이 필요하다고 느껴 쉼 없이 빵을 만들고 있을 때 제 빵 동기가 비건 프레첼을 만들고 싶다고 했고, 그 말에 "내가 만들어볼게"라고 나선 것이 제 비건 베이킹의 시작이었죠. 비건 프레첼을 만든 후, 비건 크루아상도 만들어보고 싶다는 단순한 호기심이 지금의 VAKE를 있게 했어요.

기존 비건 크루아상은 식물성 마가린으로 만든 게 전부라 코코넛 오일로 만든 크루아상은 어느 곳에도 자료가 없는 탓에 도움받을 곳이 없었어요. 그래서 무작정 만들고 생각나는 대로 빼보고, 더해보면서 만들 수밖에 없었죠. 처음부터 아예 실패했으면 시도하지 않았을 텐데, 될 듯 말 듯 안 되니 약이 올라 포기도 할 수 없었어요. 살면서 이렇게 간절히 원하고 노력한 적이 있었나 싶을 정도로 만들고 또 만들었죠. 학창 시절에 이런 식으로 공부를 했으면 뭐가 되어도 됐겠단 생각을 수도 없이 했어요.

백승도

우리나라에서 비건은 건강, 유기농, 다이어트식으로 알고 계신 분이 많던데, 저는 유제품 섭취에 어려움을 느끼는 분이나 동물 복지를 실천하는 분 등 엄격한 채식주의자도 먹을 수 있고 논비건이 먹어도 맛있는 빵을 만드는 것을 목표로 삼고 있어요.

VAKE에서는 비건 빵뿐 아니라 논비건 빵도 판매하다 보니 할 일이 두 배나 많아요. 몸은 무척 힘들지만,일산의 12평짜리 작은 빵집에 늘 찾아와주시는 동네 주민분들, 먼 곳에서 마음먹고 찾아와주시는 고마운 분들 덕분에 너무나 행복하게 보내고 있어요.

VAKE의 빵이 더 많이 알려질 수 있도록 최선을 다해 더욱더 노력하겠습니다. 이 책을 출간하고 VAKE를 오픈하기까지 물심양면으로 도움을 준 제 빵 친구들과 사랑하는 우리 가족, 너무 감사합니다. 건강 잘 지키며 행복하게 살아요, 우리!

I ran a class called 'Kiss the cake' for more than 10 years, and taught how to make butter cream flower cakes. I loved making cakes--the flowers made from butter cream blooming from the pastry bags were a scene to behold. My devotion to making the cakes was so much that I wanted to keep going them until the end of time.

But a few years ago, I had a chance to take baking class in France. Even though it lacked quality equipment and facilities, I loved the French artisan's passion for bread. The baguettes, magically rising in the oven, captivated my heart. This experience was enough to make me immediately take INBP French bread class back in Korea.

I simply thought that I would be able to fully understand baking once I finish the course. However, it was a gross understatement. Even with the same recipe, the results varied depending on the baking environment. Sometimes, I couldn't figure out the cause of the unsatisfying results. But ironically, this frustration is the source of my passion. I feel alive whenever I am studying breads, and I look forward to my upcoming days.

One day, my colleague said that she wants to make a vegan pretzels, and I told her that I will that was the beginning of my vegan baking. After making vegan pretzels, I tried to make vegan croissants. That simple curiosity is the cornerstone for 'VAKE' today.

The existing vegan croissants were made of vegan margarine, and there was no data for using coconut oil. What I could do was to keep trying with another recipe with a slight variation. If it didn't work from the beginning, I wouldn't have tried. But I couldn't give up because it seemed like it would work.

Many Koreans think that 'vegan' is healthy, organic, or for a diet. However, my vegan bread is for those who have difficulty with dairy products or for those who practice animal welfare. My goal is to make delicious vegan breads that non-vegans can enjoy together.

As 'VAKE' sells vegan and non-vegan breads at the same time, work is doubled. Even though hard work makes me exhausted, I am happy thanks to the local residents and visitors from afar who drop by my small bakery and enjoy my bread.

I would like to thank my beloved friends and family for helping me to not only open my bakery 'VAKE', but also to be able to publish this book. I hope every one of you stay healthy and happy!

CONTENTS

작가의 말 · 004

INTRO · 008

비건 베이킹에 필요한 도구 · 010

PART 1	PART 2	PART 3
매일 먹어도 부담 없는 데일리 빵	상상 이상의 달콤함, 간식 빵	한 가지 반죽으로 만드는 세가지 빵

PART 1

플레인 식빵 · 038
VEGAN PLAIN BREAD

녹차 크랜베리 식빵 · 047
VEGAN MATCHA CRANBERRY BREAD

가나슈 식빵 · 056
VEGAN GANACHE BREAD

단호박 식빵 · 065
VEGAN SUGAR PUMPKIN BREAD

올리브 베이글 · 074
VEGAN OLIVE BAGEL

호두 크랜베리 베이글 · 083
VEGAN WALNUT CRANBERRY BAGEL

시금치 치아바타 · 092
VEGAN SPINACH CIABATTA

올리브 치아바타 · 101
VEGAN OLIVE CIABATTA

호두 크랜베리 깜빠뉴 · 110
WALNUT CRANBERRY
CAMPAGNE BREAD

PART 2

초콜릿 빵 · 122
VEGAN CHOCOLATE BREAD

도넛 · 131
VEGAN DOUGHNUTS

브라우니 · 140
VEGAN BROWNIES

PART 3

세 가지 빵을 완성하는
만능 반죽 · 148
HOW TO ENJOY 3 KINDS OF BREAD
WITH A SINGLE DOUGH

단팥빵 · 152
VEGAN SWEET RED-BEAN BREAD
'DAN PAT BBANG'

맘모스빵 · 159
VEGAN MAMMOTH BREAD

인절미크림빵 · 166
VEGAN INJEOLMI CREAM BREAD

비건 베이킹에 필요한 재료 · 016

만능 비건 버터와 소스 · 024

비건 베이킹 시작 전 꼭 알아야 할 주의 사항 · 035

PART 4

비건 크루아상 반죽으로
만드는 여섯 가지 빵

비건 크루아상 반죽 · 174
VEGAN CROISSANT DOUGH

크루아상 · 184
VEGAN CROISSANT

크러핀 · 191
VEGAN CRUFFIN

빵 오 쇼콜라 · 198
VEGAN PAIN AU CHOCOLAT

빵 오 크랜베리 · 205
VEGAN PAIN AUX CRANBERRIES

올리브 타프나드 &
튀긴 양파 페스츄리 · 212
VEGAN OLIVE TAPENADE &
FRIED ONION PASTRIES

과일 & 비건 크림 페스츄리 · 219
FRUIT & VEGAN CREAM PASTRY

PART 5

주말 아침을 특별하게
비건 홈브런치

비건 햄버거 · 228
VEGAN BURGER

비건 샌드위치 · 240
VEGAN SANDWICH

비건 요거트 · 244
VEGAN YOGURT

INTRO_

더 맛있는 비건 베이킹을 시작해보세요!

비건 베이킹에
필요한 도구

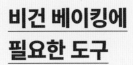

제빵에는 제과에 비해 다양한 장비와 도구가 필요하다. 저울, 온도계, 스크래퍼, 오븐 등 필수 도구는 인터넷에서 쉽게 구입할 수 있고, 여유가 된다면 가정용 믹싱기도 있으면 좋다. 처음부터 값비싼 도구를 사는 것보다는 최소한의 도구와 장비로 시도해보는 것이 좋겠다.

Bread requires a variety of equipment and tools compared to confectionery.
Essential tools such as scales, thermometers, scrapers, and ovens can be easily purchased on the Internet, and if you can afford it, it is good to have a household mixer.
But do note that rather than buying expensive tools from the start, it's better to try minimally.

Tools needed
for bread

① 간이 발효 박스

26°C가 넘지 않게 주의한다. 반죽의 온도에 따라 1차 발효 시간이 변동될 수 있다. 반죽 온도에 따라 이스트의 발효력이 달라지므로 상황에 맞게 조정해야 반죽마다 일정한 발효 상태를 맞출 수 있다. 반죽 최종 온도가 높으면 1차 발효 시간을 다소 줄이고 서늘한 곳에서 발효한다.

1. Simple fermentation box (proofing box)

When fermenting bread at home, it is usually fermented with hot water in the oven. If you have two ovens, one will be used for fermentation, and the other will be used for baking. However, we usually don't have two ovens at home, and that's why I recommend buying one of these. You can buy a 72L sealed box on the Internet and stack them like a fermenting machine.

재료 : 리빙 박스, 오븐 팬, 목봉(종이컵으로 대체 가능), 두꺼운 비닐(김장 비닐), 온습도계, 뜨거운 물, 컵

① **리빙 박스 :** 인터넷 사이트에서 구입할 수 있으면 72L 이상의 사이즈를 구입한다. 오븐 팬과 목봉을 조립해서 발효기처럼 사용할 수 있다.

② **오븐 팬 :** 여러 개 있으면 많은 개수의 단팥빵이나 베이글을 발효하는 데 유용하다.

③ **목봉 :** 식빵같이 한 배합으로 3~4개 정도 나오는 빵이라면 밀폐 용기에 넣고 습기를 유지해주면 되지만, 베이글이나 단팥빵같이 한 배합의 반죽으로 10개 이상의 반죽이 나와서 팬이 여러 개 필요한 경우라면 목봉을 팬 끝 쪽에 놓고 팬을 올리면 4-5개의 팬도 무리 없이 발효할 때 사용 가능하다. 없다면 인터넷 포털사이트에서 재단 요청을 하거나, 종이컵으로 대신해도 무방하다.

④ **온습도계 :** 간이 발효 박스를 만들어서 사용할 경우 온습도계를 넣어두면 쉽게 온도와 습도를 확인할 수 있으므로 편하다.

Materials needed : Living box, oven pan, wooden rolling pin (replaceable with paper cups), thick vinyl (kimchi vinyl), thermo-hygrometer, hot water, cup

① **Living Box :** If you can buy a size that has a capacity of 72L or more, you can assemble an oven pan and a wood stick and use it like a fermenter.

② **Oven pan :** Having several pans is useful for fermenting a large number of sweet red bean bread or bagels.

② **Wooden rolling pin (non-movable) :** If you have 3 to 4 breads from one dough, you can put them in a sealed container and maintain moisture. If you need more than 10 doughs at once, such as bagels or sweet red-bean breads, place the rolling pin at the end of the pan and place the pan on top. You will be able to use 4~5 pans when fermenting. If this is not possible for you, use paper cups as a replacement. You can try requesting for the dough to be cut when buying instead as well.

② **Thermo-hygrometer :** When making and using a simple fermentation box, this is convenient because you can easily check the temperature and humidity.

만드는 과정

① 팬을 놓고 네 모서리 끝에 목봉이나 종이컵을 놓는다. 팬 중간 부분에 낮은 그릇에 뜨거운 물을 넣어 리빙 박스의 온도와 습도를 올려준다.

② 빵 성형이 끝나면 온습도를 맞춘 리빙 박스에 빵 반죽을 넣고 발효시킨다.

How to make the DIY fermentation box

① Place the rolling pin or paper cups at the end of the four corners of the pan. Place hot water in a low bowl in the middle of the pan to increase the temperature and humidity of the box.

② After the bread molding is completed, place the bread dough in the box for fermentation to occur.

② 전자저울

베이킹을 할 때는 정확한 계량이 필요하다. 특히 소금이나 이스트의 경우는 양이 조금만 차이나도 차거나 발효가 많이 진행되거나 하는 경우가 생기기 때문에 저울은 필수품이다.

2. Electronic scale

Accurate measurements are needed when baking. The scale is a necessity in the case of salt or yeast due to high possibilities of variations of the end result.

③ 베이킹용 온도계

빵 반죽을 할 때 최종 온도는 1차 발효나 2차 발효 시간을 정하는 기준이 되므로 온도계를 준비해서 체크해서 반죽 온도가 높다면 1차 발효 시간을 줄이고, 반죽 온도가 너무 낮다면 발효 온도에 신경 쓰고 시간을 늘리는 등 조절해줘야 하므로 온도계는 필수 도구다.

3. Thermometer for baking

When kneading bread, the final temperature is a criterion for determining the time of primary or secondary fermentation. Prepare a thermometer and pay close attention to it in order to reduce or increase the first fermentation time if the temperature of the dough is too high or low respectively.

④ 스크래퍼

반죽을 볼에서 떼어내거나 분할할 때 유용한 도구다. 둥근 것은 믹싱기에서 반죽을 떼어낼 때, 일자는 반죽을 분할할 때 등 용도에 맞게 사용하면 된다.

4. Scraper

This is a useful tool when removing or dividing dough from a bowl. Round ones can be used for removing the dough from the mixer, and straight ones can be used for dividing the dough.

⑤ 나무판

크루아상 작업을 할 때 버터가 많이 들어가기 때문에 냉동실에 넣어두고 차갑게 한 뒤 반죽을 올려 작업을 하면 반죽이 수축되는걸 막아줘서 유용하고, 크루아상을 재단할 때도 칼을 이용해서 자르기도 편하다.

5. Wooden board

Put the board in the freezer to cool it, and take it out when making Croissants. Place the dough on the cold board to prevent the dough from shrinking (This is due to lots of butter being used in making Croissants). You can also easily use knives to cut the Croissant on the board as well.

⑥ 체

제과 제품을 만들 때 말차가루나 코코아 파우더, 콩가루 등 뭉쳐 있는 가루를 체에 내린 후 사용하는 것이 좋다. 뭉친 가루를 풀어주는 것만이 아니라, 체를 쳐주면 가루에 공기가 들어가 더 부드러운 제품을 만들 수 있다.

6. Sieve

When making confectionery products, it is recommended to sift lumped powders such as green tea powder, cocoa powder, and bean powder. Not only does sifting loosen the lumped up powder, but air can enter the powder and make it softer.

⑦ 밀대

단팥빵, 식빵 등의 반죽 성형 시, 또는 크루아상 반죽을 균일하게 밀어야 할 때 사용하는 필수품이다. 균일한 두께로 밀어 비슷한 크기로 나누면 원하는 무게로 분할하기가 쉽다. 반죽을 분할할 때는 자투리 여러 조각을 뭉쳐놓는 것보다 한 덩어리의 반죽으로 분할하는 것이 좋다.

7. Roller

Rollers are for molding dough evenly, and is a requirement for many breads such as Croissants and Sweet red-bean breads. It becomes much easier to divide the dough into the desired weight if you roll the dough with the roller into a uniform thickness, and divide it into similar sizes. When dividing the dough, it is better to divide it into a large lump of dough than lumping several pieces together.

⑧ 실리콘 주걱

재료를 섞을 때 뿐 아니라 섞어놓은 재료를 깔끔하게 긁어서 사용하기 때문에 재료를 버리지 않고 모두 사용할 수 있다.

8. Silicone spatula

Great for mixing and scraping the remaining ingredients from the mixer or bowls.

⑨ 거품기

반죽을 가볍게 섞을 때, 비건 크림을 끓일 경우 크림이 매끄럽게 끓을 수 있게하는 데 사용하면 좋다.

9. Whisk

Use this when mixing the dough lightly, or to stir vegan cream when boiling it.

⑩ 짤주머니

도넛에 크림을 주입할 때 사용하면 좋다.

10. A piping bag

Used for injecting cream into donuts.

⑪ 테프론 시트

코팅되어 있는 철판이라면 사용하지 않아도 되지만, 논코팅 철판이라면 테프론 시트를 깔고 단팥빵이나 쿠키, 스콘 등을 구운 후 떼어낼 때 쉽게 떼어낼 수 있어서 유용하다. 하드 빵을 구울 때도 달군 돌판에 테프론 시트에 얹은 반죽을 올려 돌판에 밀어넣어 구울 수 있다.

11. Teflon sheet

If your iron pan is coated, you won't need this. This is used for cookies, scones, sweet red-bean breads and more to avoid sticking to the pan. You can also use this on stone plates when baking hard bread.

⑫ 바게트천

깜빠뉴나 치아바타를 성형하고 발효를 할 때 천을 이용해 벽을 만들어서 발효를 조절하는 용도로 사용한다.

12. Baguette cloth

Control the fermentation process of Campagne or Ciabatta by making a wall with this cloth.

⑬ 쿠프 칼

깜빠뉴에 쿠프를 내는 용도로 사용한다.

13. Baker's Blade

Used to cut slits in hard bread, such as Campagne and Baguettes.

⑭ 식힘망

오븐에서 빵이 나오면 틀에서 바로 빼내 식힘망에 올려서 식혀야 눅눅해지지 않고 잘 건조시킬 수 있다.

14. Cooling net

In order to avoid making the bread soggy, place the bread immediately into a cooling net to dry when it comes out of the oven.

⑮ 목란

단팥빵 성형 후 2차 발효 전 평평해진 반죽의 중심 부분을 눌러주는 용도로 사용하기 유용한 도구다. 도넛을 만들 때 둥글리기 한 반죽을 평평하게 눌러줄 때도 사용하면 균일하게 눌러주는 데 도움을 준다.

15. Wooden mold

This tool is used to mold bread shapes. For an example, before fermenting the sweet red-bean bread a second time, press the mold onto the dough to make a large dimple in the middle. The mold is also used to flatten donut dough as well.

⑯ 사각틀, 식빵틀

원하는 품목에 따라 틀을 선택해 알맞은 용도로 사용한다.

16. Square frames, bread molds / frames

Choose a frame shape according to your desired bread.

⑰ 큰 칼

크루아상, 뺑 오 쇼콜라, 파이 반죽을 재단할 때 한번에 자를 수 있으므로 유용하다.

17. Large knife

Useful in cutting Croissant, Pain au Chocolat, and Pie dough in one slice.

⑱ 오븐 팬

사용하는 오븐마다 적합한 오븐 팬이 있는데, 코팅된 팬이 사용하기 편하다.

18. Oven pan

Each oven has a suitable oven pan, but using a coated pan is much more convenient.

19. 붓

부드러운 털 붓이 사용하기 편하며 달걀이나 기름을 바를 때 또는 밀가루를 터는 용도로 사용하면 편하다.

19. Brush

Brushes with soft hair are great when applying eggs or oil to flour. Use this as well when brushing flour off of bread.

20. 타이머

1분, 5분, 10분 단위 모두 세팅할 수 있는 타이머가 사용하기 편하다.

20. Timer

A timer that can be set in 1 minute, 5 minutes, and 10 minutes intervals is recommended.

21. 밀폐 용기

믹싱이 끝난 반죽을 발효하는 작은 통이면 된다. 반죽 양이 많지 않은데, 통의 면적이 너무 넓다면 반죽이 퍼져서 발효가 제대로 되지 않으므로 반죽 양에 맞는 사이즈의 밀폐 용기를 준비하는 것이 좋다.

21. Sealed container

Needed to ferment the mixed dough. Prepare a container size according to the amount of dough, as the dough cannot ferment properly if the container is too large.

22. 스테인레스 볼

비건 크림을 만들 때 재료를 섞는 용도 등 여러모로 사용하면 좋다.

22. Stainless steel bowl

Needed to mix ingredients while making vegan cream.

23. 면적이 넓은 냄비

도넛을 튀기거나 베이글을 데치는 용도로 사용한다. 가스레인지에 올릴 수 있는 스테인 볼을 사용해도 좋다.

23. A large pot

Needed to deep fry donuts or to slightly parboil bagels. A stainless bowl that can be put on the gas stove can also be used.

24. 체

베이글을 건져낼 수 있는 평평한 체를 준비하면 베이글이나 도넛 튀기기에 좋다.

24. Sieve

Prepare a flat sieve in advance for the the bagels or donuts to rest after frying.

25. 블렌더

비건 버터를 만들 때 아주 유용하게 사용하는 필수품이다.

25. Blender

Useful in general, and necessary to make vegan butter.

비건 베이킹에
필요한 재료

INTRO 2

논비건이 먹어도 맛있는 비건 빵을 만들고 싶었어요. 우리나라에서 비건 빵은 건강, 유기농, 다이어트의 인식이 강한 것 같아요. 달걀, 유제품, 젤라틴 등 동물성 재료를 완전히 배제하고 식물성 재료만을 사용해 만드는, 누가 먹어도 맛있는 빵을 만드는데 초점을 두었어요. 특정 동물성 재료에 알러지가 있는 분, 신념으로 비건을 실천하시는 분 등 비건이 마음놓고 먹을 수 있는 식물성 재료만으로 맛있는 빵을 만들어보실까요?

I wanted to make vegan bread that even non-vegans can enjoy. In Korea, people think vegan breads are healthy, organic, or for diet. Therefore, I focused on making breads that are delicious for anyone, using vegetable ingredients only and excluding animal ingredients such as eggs, dairy products, and gelatin. For those who are allergic to some animal ingredients, and those who practice veganism with their beliefs, Shall we make delicious bread with vegetable ingredients that Vegans can enjoy?

Ingredients
for vegan bread

① 강력분

글루텐 함량이 많은 밀가루로 단백질 함량이 13% 이상인 강력 밀을 제분해 얻으며 주로 빵을 만드는 데 사용한다.

1. Strong flour(Hard flour)

Mainly used to make bread. Made by milling strong wheat that contains 13% or more protein.

② 중력분

강력분과 박력분의 중간 정도의 밀가루다. 단백질 함량은 10%. 빵은 무조건 강력분으로 만들어야 한다고 생각하기 쉽지만, 원하는 제품에 따라 중력분을 사용해도 좋다.

2. Plain flour

This is in the middle between strong and weak flour. Protein content is around 10%. It's commonly thought by many that bread should be made of strong flour, but you can also use plain flour depending on the bread.

③ 박력분

보통 제과 제품을 만들 때 많이 사용한다. 단백질 함량은 9%. 빵은 강력분으로만 만든다고 생각하기 쉬운데, 쫄깃함뿐 아니라 부드러움도 같이 주는 경우 박력분을 섞어서 만들기도 한다.

3. Weak flour(Soft flour)

Usually used in confectionery. Protein content is around 9%. Soft flour can be mixed with strong flour when you want that chewy and soft texture in breads.

④ 트레디션 밀가루

프랑스 밀가루로, 우리나라와 달리 type으로 표시한다. 빵을 만들 때는 T65 또는 T55를 사용한다. T는 type의 약자다. 깜빠뉴와 치아바타 만들 때 주로 사용하는데, 가정에서 가정용 컨벡션으로 하드 빵을 굽는 것이 쉬운 것만은 아니므로 이 책에서는 강력분을 섞어서 사용한다.

4. Tradition flour

French flour, which is marked in type. T65 or T55 is used to make bread.('T' stands for type.) It is mainly used to make 'Pain de campagne' and 'Ciabatta'. As it is not easy to bake hard breads with small convection oven at home, I suggest to mix strong flour with tradition flour in this book.

⑤ 통밀가루

정제하지 않은 밀로 섬유질, 비타민, 무기질 등 영양학적으로 우수하고, 식이 섬유가 풍부하며, 구수한 맛이 좋지만, 밀가루보다 식감이 거칠다.

5. Whole wheat flour

Unrefined wheat. Nutritionally excellent. Rich in dietary fiber and vitamin. Great savory taste, but has a rougher texture than regular flour.

⑥ 호밀가루

호밀은 글루텐 함량이 적어 밀가루와 섞어서 빵을 만들기 쉬워 보통 프랑스 시골빵인 깜빠뉴를 만들 때 사용한다. 요즘은 천연 발효빵에 관심이 많아져 르방을 이용해서 호밀만으로 호밀빵을 만들기도 한다.

6. Rye flour

Rye has a low gluten content, so it's easy to mix it with flour to make bread. It's usually used to make French country bread, "Pain de Campagne". Nowadays, as people are more interested in naturally fermented bread, rye bread can be made simply by using levain on rye.

⑦ 두유

우유 대신 무난하게 사용할수 있는 재료다. 비건을 실천하는 사람뿐 아니라 유당불내증이 있는 사람들에게도 좋은 재료다.

7. Soy milk

Can substitute milk. Good for vegans and for those who have lactose intolerance.

⑧ 아몬드 밀크

두유와 마찬가지로 우유 대신 사용할 수 있는데, 특유의 향이 있다.

8. Almond Milk

Can substitute milk. Has a unique scent, unlike soy milk.

⑨ 코코넛 밀크

코코넛 워터가 어린 코코넛을 활용하는 것에 비해 코코넛 밀크는 성숙한 코코넛을 활용하는 것이 보통이다. 빵을 만들 때 우유 대신 넣을 수 있고, 비건 요거트를 만들수 있다.

9. Coconut Milk

Mature coconuts are used to make coconut milk, whereas young coconuts are used to make coconut water. Coconut can substitute milk in baking, and can make vegan yogurt.

⑩ 코코넛 크림

코코넛 크림은 코코넛 밀크와 비슷하지만 더 진하고 지방 함량이 높다. 아이스크림이나 달콤한 디저트에 많이 쓰며 비건 휘핑크림을 만드는 데 유용한 재료다.

10. Coconut cream

Similar to coconut milk, but thicker and has a higher fat content. It is used for ice cream and sweet dessert dishes and is a useful ingredient for making vegan whipped cream.

⑪ 과일 퓨레

과일을 갈거나 으깨어 만들어놓은 것으로 과일잼을 만들어 빵을 만들 때 사용하면 좋다.

11. Fruit puree

Made by grinding or crushing fruit. If you make fruit cream with this it is useful to making bread.

⑫ 전분

보통 옥수수 전분을 사용하며 과일 퓨레나 크림 제조 시 사용한다.

12. Starch

Corn starch is usually used when making fruit purees or cream.

⑬ 올리브유

올리브 열매에서 추출한 기름으로 엑스트라 버진 올리브유는 과육을 냉압해 얻고, 빵을 찍어 먹는 등 가열하지 않는 요리나 드레싱용으로 사용하면 좋고, 퓨어 올리브유는 식용유를 대체해서 튀기거나 볶는 데 사용하면 좋다. 이 책에서는 치아바타와 푸가스를 만들 때 사용하므로 엑스트라 버진 올리브유를 사용한다.

13. Olive Oil

Extracted from olive fruit. Extra virgin olive oil is made by cold pressure on the fruit and is good for dipping bread or salad dressing, while pure olive oil is good for frying. In this book, extra virgin olive oil is used to make Ciabatta and Fougasse.

⑭ 식물성 오일

냄새가 나지 않는 식물성 오일을 사용하면 된다. 도넛을 튀기거나 충전용 크림을 만들 때 사용한다.

14. Vegetable oil

Use vegetable oils that do not have any scent. Use this to fry doughnuts or to make filling cream.

⑮ 호두 오일

호두를 볶아서 압착해 만들기 때문에 가벼운 호두 향이 나므로 빵 반죽의 글루텐이 생긴 후 호두를 넣을 때 오일을 같이 넣으면 빵에 풍미를 더한다. 개봉 후에는 냉장고에 보관하며 빠른 시간 내에 사용하는 것이 좋다.

15. Walnut oil

Has a light walnut smell, as it is made by stir-frying and pressing walnuts. Adding walnuts and walnut oil after gluten is formed in the dough enhances the flavor. It is recommended to store it in the refrigerator after opening, and to use it as soon as possible.

⑯ 비정제 설탕

정제 과정을 거치지 않아서 비타민, 무기질 미네랄 성분이 남아 있는 설탕이며 정제한 백설탕에 당밀을 첨가해 만드는 황설탕, 흑설탕과는 다르다.

16. unrefined sugar

Unlike refined sugar, this still contains the natural vitamins and minerals. Note that brown or black sugar is made by adding molasses syrup to refined sugar.

17 설탕

사탕수수 줄기에서 정제하는 것으로 비타민이나 미네랄이 없고, 단맛을 낸다. 제과나 제빵에서 설탕은 단맛만 내는 것이 아니라 발효를 촉진하고 빵을 부드럽게 하는 역할도 하니 단맛이 싫다고 무조건 설탕을 줄이는 것은 바람직하지 않다. 정제 설탕(백설탕)은 태운 동물의 뼈로 만든 탄화 골분을 이용해서 정제한다. 황설탕도 이 정제된 설탕에 당밀을 첨가하는 방식으로 만들어진다.

17. Sugar

Refined from sugar cane stems, and has no vitamins or minerals. As sugar promotes fermentation and softens bread, it is not recommended to reduce sugar in order to reduce the sweetness. When making refined sugar from sugar canes, the refining process is done through using the carbonized bone powder of burned animals. Brown sugars can be made by adding molasses to refined sugar.

18 슈가 파우더

백설탕을 곱게 분쇄한 설탕으로 전분을 3~5% 첨가한 설탕을 일컫는다. 액체에도, 버터에도 잘 녹고 잘 섞이기 때문에 작업하기 쉽다.

18. Sugar Powder

White sugar is finely pulverized sugar with 3-5% starch added. It is easy to use as it melts well in liquids and butter.

19 당밀

사탕수수를 설탕으로 가공하는 과정에서 남은, 결정화되지 않은 부산물로 어두운 갈색의 액체이며, 철분, 칼슘, 칼륨과 비타민 B6, 구리, 셀레늄, 망간 등 다양한 영양소를 함유하고 있다. 빵의 발효에 도움을 준다.

19. Molasses

Dark brown liquid left when processing sugar cane into sugar. An uncrystallized by-product, and contains various nutrients such as iron, calcium, potassium, and vitamin B6, copper, selenium, and manganese. This helps in bread fermentation.

20 소금

소금을 첨가하지 않은 빵 반죽은 밀가루 특유의 구수하고 달콤한 풍미를 제대로 내지 못한다. 소금은 단순히 짠맛을 내는 역할을 하는 것이 아니라 다른 풍미도 이끌어내고 글루텐 구조를 더 강하게 만든다. 소금을 넣지 않고 반죽을 치면 빵 볼륨도 작고, 발효 속도와 빵의 구움 색에도 영향을 미친다. 믹싱기로 반죽을 치는 동안 소금이 녹아야 하므로 입자가 가는 소금을 사용하는 것이 좋다.

20. Salt

Bread dough without salt does not give off the savory and sweet flavor of flour. Salt brings out savory and makes the gluten structure stronger. If you knead without salt, the volume of the bread gets smaller and the fermentation speed and baking color of the bread are affected. It is recommended to use fine salt because the salt must melt while kneading with a mixer.

21 물

물은 빵 반죽에서 중요한 재료다. 밀가루와 섞였을 때 글루텐을 만들고 소금, 설탕, 이스트 같은 재료를 녹이며 이스트의 발효와 번식에도 필요하다.

21. Water

Water is important in bread dough. Water makes gluten when mixed with flour, and dissolves salt, sugar, and yeast. Water is also needed for yeast fermentation and reproduction.

22 이스트

다량의 이산화탄소를 발생시켜 빵을 부풀게 하는 작용을 한다. 생이스트를 사용하는 것이 가장 좋지만, 생이스트는 유통기간이 짧아 가정에서는 기간 내에 사용하지 못할 수 있으니 세미 드라이 이스트를 사용하면 된다. 실온에 보관 가능한 제품도 있는데, 보관이 편하다는 장점이 있지만,

발효 시 막걸리 냄새가 난다는 평이 많다. 설탕이나 버터가 많이 들어가는 크루아상, 단팥빵, 식빵은 골드 드라이이스트, 깜빠뉴나 치아바타 같은 하드 빵은 레드 드라이이스트를 사용하면 된다.

22. Yeast

Raises the bread dough as it generates a large amount of carbon dioxide. It is best to use raw yeast, but as raw yeast has a short shelf life, you can use semi-dry yeast at home. There are some products that can be stored at room temperature. Although they can be easily stored, many say that they smell like "Makgeolli" (makgeolli: Korean traditional liquor based on rice) during fermentation. Use gold dry yeast for Croissants, Sweet red-bean bread, and bread that contain a lot of sugar or butter. Use red dry yeast for hard bread such as Pain de campagne or Ciabatta.

(23) **통팥앙금**

팥을 삶아 설탕을 넣어 조려서 만든 것으로 설탕의 비율로 단맛 정도를 조절한다. 시중 제품에도 저당 팥앙금 이 있으니 선택해서 구매한다.

23. Red-bean paste

Boiled red-bean paste with sugar. The sweetness can be adjusted by the sugar amount. You can find low-sugar red-bean paste from the grocery store.

(24) **호두**

단팥빵, 깜빠뉴를 만들 때 사용한다. 견과류는 그냥 사용하는 것보다 구워서 사용하면 견과류의 고소한 맛이 더 살아나기 때문에 사용 전 구워서 식힌 후 사용하는 것이 좋다. 껍질을 깐 호두는 산패되기 쉽기 때문에 밀봉해 냉동 보관하는 것이 좋다.

24. Walnut

Used for Sweet red-bean bread and Pain de Campagne. Nuts are better to be baked before use to enhance its savory taste. Peeled walnuts are recommended to be kept frozen, as they can easily be acidified.

(25) **아가베 시럽**

용설란에서 추출한 당분을 이용해 만든 시럽. 설탕 대신 사용할 수 있다.

25. Agave Syrup

Syrup made by sugar extracted from the Agave. It can substitute sugar.

(26) **메이플 시럽**

메이플 나무의 수액에서 채취해 만드는 천연당으로 특유의 독특한 향을 지니고 있다.

26. Maple Syrup

A natural sugar that is harvested from the sap of maple trees. It has a unique scent.

(27) **말차가루**

1차 가공한 녹차를 분쇄해 만든 가루 녹차. 빵을 만들 때 너무 많이 넣으면 쓴맛이 날 수 있으니 밀가루 대비 5%를 넘지 않도록 한다.

27. Matcha powder

Powdered green tea made by grinding processed green tea. To use less than 5% of the flour is recommended, as it can taste bitter if you add too much.

(28) **코코넛 오일**

24℃ 이하에서는 고체, 그 이상일 때는 액체 상태로 바뀌는 오일. 코코넛 특유의 향을 내므로 원하지 않는다면 정제된 코코넛 오일을 사용하면 된다.

28. Coconut oil

It changes to a solid state under 24℃(75°F) and to a liquid state above that temperature. If you don't like the coconut scent, you can also use refined coconut oil.

(29) **바닐라 익스트랙**

초콜릿, 아이스크림, 케이크, 과자, 크림 등에 널리 사용하며 바닐라 꼬투리에 에탄올을 담가 만든다. 바닐라빈을 사용하는 것이 가장 좋지만, 바닐라 익스트랙을 사용해도 무방하다.

29. Vanilla Extract

Widely used in chocolate, ice cream, cakes, snacks, cream, etc. Made by immersing ethanol in vanilla pods. It is best to use vanilla bean, but you can try vanilla extract instead.

(30) 카카오 파우더

볶은 카카오를 분쇄해서 가루로 만든 다음 압착해 카카오 버터를 분리하고 그 나머지를 건조·분쇄한 가루다. 설탕이나 첨가물을 섞지 않은 가루를 사용하는 것이 좋다.

30. Cacao Powder

Made by pulverizing and pressing the roasted cacao to separate cacao butter. It is recommended to use cacao powder that contains no sugar or other additives.

(31) 콩가루

볶거나 구운 콩가루를 사용한다. 생콩가루가 있다면 오븐에 구워서 사용해도 좋다. 콩가루는 잘 뭉치므로 사용할 때는 체를 쳐서 이용한다.

31. Soybean powder

Use roasted or baked powder. If you have some raw soybean powder, you can bake it in an oven before use. Use it through a sieve as this clumps well.

(32) 쑥가루

쑥을 말린 후 곱게 빻아 만든 가루. 비건 크림, 제과에 이용한다.

32. Mugwort powder

Made by drying and grinding mugwort. It is used for vegan cream and confectionery.

(33) 다크 초콜릿

성분은 카카오매스와 설탕, 바닐라, 레시틴정도다. 밀크 초콜릿과 화이트 초콜릿은 카카오에 분유, 연유, 우유 등을 넣어 가공한 초콜릿이어서 비건 제품을 만들기 힘든 데 비해 다크 초콜릿은 유제품을 함유하지 않았다.

33. Dark chocolate

Ingredients of dark chocolate are cacao mass, sugar, vanilla, and lecithin. Dark chocolate does not contain dairy products, whereas milk chocolate and white chocolate are processed chocolate with powdered or condensed milk, and then difficult to make vegan products.

(34) 건조 무화과

말린 무화과. 식이 섬유와 철, 칼슘, 포타슘, 마그네슘과 폴리페놀 함량도 높다. 반죽에 넣어 빵을 만들 때는 전 처리를 해서 사용하는 것이 좋다.

34. Dried figs

Contains dietary fibers, iron, calcium, potassium, magnesium and polyphenol. It is recommended to pre-process the bread dough.

(35) 아몬드 파우더

밀가루 대신 사용하거나 비건 파마산 치즈에 사용한다.

35. Almond powder

Can substitute flour. Used for vegan parmesan cheese.

(36) 애플 사이다 비네거

천연 발효 식초. 비건 버터를 만들 때 사용하는 필수 재료다.

36. Apple Cider Vinegar

Natural fermented vinegar. It is an essential ingredient to make vegan butter.

(37) 냉동 라즈베리

브라우니 반죽에 냉동 상태로 넣어서 굽거나 잼으로 활용해 빵에 충전하기 쉽다.

37. Frozen raspberry

Can be put in brownie dough, or used as bread filling in the form of raspberry jam.

③⑧ 아마씨 가루

비건 달걀액을 만들 때 물과 섞어서 사용한다.

38. Flax seed powder

Mix with water when making vegan egg solution.

③⑨ 올리브

푸가스나 치아바타, 포카치아, 베이글 등 대부분의 빵을 만들 때 사용할 수 있다. 그린 올리브, 블랙 올리브 등을 선택해 사용하거나 같이 사용해도 좋다.

39. Olive

Can be used to make most breads including Fougasse, Ciabatta, Focaccia, and Bagels. You can choose one between green olives and black olives, or use both.

④⓪ 양파가루

비건 파마산 치즈를 만들 때 소량 이용한다.

40. Onion powder

Small amount is used for vegan parmesan cheese.

④① 후추

비건 베이글을 만들 때 사용한다. 빵과 안 어울릴 것 같지만, 후추 향과 베이글이 무척 잘 어울린다.

41. Pepper

Used for vegan bagels. The scent of pepper and bagels go very well together.

④② 영양 효모

주로 당밀에서 배양되며 필수아미노산과 다양한 비타민 B군을 함유하고 있다. 비건 파마산 치즈, 비건 버터, 비건 치즈 만들 때 두루두루 사용한다.

42. Nutritional yeast

Mainly cultured in molasses and contains essential amino acids and various vitamin B groups. Used to make vegan parmesan cheese, vegan butter, and vegan cheese.

④③ 조청

각종 곡물의 전분질에 엿기름을 넣고 뭉근히 고아 발효시켜 만들어 꿀 대신 쓰면 좋다. 요즘은 주로 쌀로 만들고 진한 갈색이 돌며 농도가 되직하다.

43. Grain syrup

Can substitute honey. Made by adding malt oil to the starch of various grains and fermenting them. Usually made of rice nowadays, and has a dark brown color. It has a thick concentration.

④④ 고생지

1차 발효된 사전반죽 전반죽 묵은 반죽으로 불리기도 한다. 빵을 반죽할 때 넣으면 발효를 돕고 빵의 풍미가 깊어진다. 고생지는 되도록이면 신선한 것을 사용한다.

44. Pâte fermentée

This fermented dough is mass of dough after first fermentation, also called pre-fermented dough, or old dough. This dough helps a better fermentation and give a nice taste. Use only fresh pâte fermentée if possible.

만능 비건 버터와 소스

만능 비건 버터와
소스

비건 베이킹을 시작할 때 꼭 필요한 만능 비건 버터 만드는 법
과 빵과 함께할 다양한 소스를 소개합니다.
[비건 버터 + 크림 + 잼 + 페스토 + 마요네즈 + 치즈 + 소보로]

Introducing how to make all-purpose vegan butter,
which is essential when you start vegan baking, and
a variety of sauces to go with bread.

All-purpose
vegan butter and sauces

① **비건 버터**

- 스프레드용
- 베이킹용
- 페스츄리 뚜라주용(접기용)

Vegan butter

- For spread
- For baking
- Lamination butter(Butter de tourrage)

1) 스프레드용 버터 Spread butter

재료

두유 160g

코코넛 오일 200g

소금 2g

애플 사이다 비네거 20g

설탕 10g

아몬드 파우더 100g

Ingredients

Soy milk 160g

Coconut oil 200g(liquid)

Salt 2g

Vinegar 20g

Sugar 10g

Almond powder 100g

Let's Make

1. 두유에 식초를 넣어 10초 정도 둔다.
2. 몽글거리며 점도가 생기면 나머지 재료를 넣고 핸드 블렌더로 갈아준다.

1. Add vinegar to soy milk and leave for about 10 seconds.
2. When it becomes soft and viscous, add the rest of the ingredients and blend them with a hand blender.

2) 베이킹용 버터 Baking butter

재료

두유 150g

코코넛 오일 300g

소금 1g

애플 사이다 비네거 20g

설탕 10g

뉴트리션이스트(영양 효모) 5g

Ingredients

Soy milk 150g

Coconut oil 300g(liquid)

Salt 1g

Vinegar 10g

Sugar 10g

Nutritional yeast powder 5g

Let's Make

1. 두유에 식초를 넣어 10초 정도 둔다.
2. 몽글거리며 점도가 생기면 나머지 재료를 넣고 핸드 블렌더로 갈아준다.

1. Add vinegar to soy milk and leave for about 10 seconds.
2. When it becomes soft and viscous, add the rest of the ingredients and blend them with a hand blender.

3) 페스츄리 뚜라주용(접기용) 버터　Lamination butter

재료	Ingredients
두유 100g	Soy milk 100g
액체 상태의 코코넛 오일 150g	Coconut oil 150g(liquid)
식초 10g	Vinegar 10g
소금 1g	Salt 1g
뉴트리션이스트(영양 효모) 2g	Nutritional yeast powder 2g
물엿 20g	Starch syrup 20g

Let's Make

1. 모든 재료를 넣고 블렌더로 갈아 마요네즈 상태가 되면 블렌더를 멈춘다.
2. 비닐에 140g씩 올려 감싼다.
3. 스크래퍼를 이용해 균일한 두께가 되게 한 후 사용 전까지 냉장 보관한다.

1. In blender, combine every ingredient, Blend until smooth and creamy like mayonnaise.
2. Spread the butter on a plastic wrap and enclose the butter in the plastic.
3. Shape it into a 6" square with a dough using plastic scraper and has an even thickness. Store it in refrigerator before use.

(2) **올리브 타프나드**

Olive Tapenade

재료

블랙 올리브 180g

케이퍼 20g

레몬 제스트 1개분

레몬 쥬스 20g

마늘 2알

후추 1g

드라이 토마토 30g

비건 파마산 30g

올리브유 70g

Ingredients

Black olive 180g

Caper 20g

Lemon, grated zest of 1

Lemon juice 20g

Pieces of garlic 2

Ground black pepper 1g

Dried tomatoes 30g

Vegan parmesan cheese 30g

Olive oil 70g

Let's Make

1. 모든 재료를 푸드 프로세서에 넣고 갈아준다.

1. Put all ingredients in a food processor until combined.

(3) **비건 초코 크림**

Vegan chocolate cream

재료

두유 400g

설탕 90g

소금 2g

바닐라 에센스 5g

밀가루 30g

식물성 오일 20g

다크 초콜릿 120g

Ingredients

Soy milk 400g

Sugar 90g

Salt 2g

Vanilla essence 5g

Flour 30g

Vegetable oil 20g

Dark chocolate 120g

Let's Make

1. 넓은 냄비에 다크 초콜릿을 제외한 모든 재료를 넣어 약한 불에서 저어주면서 끓인다.

2. 냄비 속 재료가 보글보글 끓어오르면 냄비를 불에서 내린 후 다크 초콜릿을 넣고 잘 섞어준다.

3. 표면이 굳지 않도록 랩으로 밀착 랩핑한 후 냉장고에 넣어서 식힌다. 완전히 식으면 짤주머니에 넣은 채로 사용 전까지 냉장 보관한다.

1. In a large saucepan, combine all cream ingredients except dark chocolate. Cook over low heat and bring to a boil, stirring often.

2. Bring to a boil, remove the pan from the heat, stir in the dark chocolate; whisk until the chocolate is melted and the mixture is smooth.

3. Place plastic wrap on the surface to prevent a skin from forming, and chill thoroughly. Spoon the cooled cream into the piping bag and refrigerate it before using.

④ 비건 인절미 크림

Vegan injeolmi cream

재료	Ingredients
두유 170g	Soy milk 170g
설탕 35g	Sugar 35g
소금 한 꼬집	Pinch of salt
바닐라빈 1/3개	Vanilla bean 1/3 pod
밀가루 12g	All purpose flour 12g
식물성 오일 8g	Vegetable oil 8g
볶은 콩가루 17g	Roasted soy bean powder 17g

 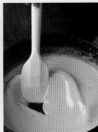  

Let's Make

1. 넓은 냄비에 볶은 콩가루를 제외한 모든 재료를 넣고 약한 불에서 저어주면서 끓인다.

2. 냄비 속 재료가 보글보글 끓어오르면 냄비를 불에서 내린 후 볶은 콩가루를 넣고 잘 섞는다.

3. 완전히 식힌 후 냉장 보관한다.

1. Add everything but roasted bean powder to a large pan over a low heat, stirring constantly and bring to the boil.

2. When it boils remove from the heat and add the roasted bean powder and mix until evenly combined.

3. Let cool completely and refrigerated.

Vegan custard cream

재료
두유 400g
설탕 80g
소금 한 꼬집
바닐라빈 1/2개
옥수수 전분 32g
식물성 오일 20g

Ingredients
Soy milk 400g
Sugar 80g
Pinch of salt
Vanilla bean 1/2 pod
Corn starch 32g
Vegetable oil 20g

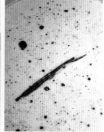

Let's Make

1. 넓은 냄비에 모든 재료를 넣고 잘 저어준다.
2. 눌어붙지 않도록 계속 저어주며 약한 불에서 끓인다.
3. 완성된 크림은 비닐 위에 붓고 마르지 않게 랩핑해 빠르게 냉각한다.

1. Put every cream ingredient in a large pan.
2. Over a low heat, stirring constantly and bring to the boil.
3. When it boils remove from the heat pour over a plastic wrap and cover the surface of cream as well. Then cooling it down quickly.

⑥ 라즈베리 잼

Raspberry jam

재료
냉동 라즈베리 과육 500g
설탕 300g

Ingredients
Frozen raspberry 500g
Sugar 300g

Let's Make

1. 냉동 라즈베리 과육에 설탕을 섞어서 즙이 많이 나오도록 냉장고에 하룻밤 동안 보관한다.

   TIP. 라즈베리가 없다면 트리플 베리나 다른 과일을 이용해도 무방하다.

2. 다음 날 체에 밭쳐 과육과 즙을 분리한 후 즙을 먼저 끓이고 졸아들면 과육을 넣어 다시 끓인다.

3. 차가운 물이 담긴 그릇에 잼을 떨어뜨려 풀어지지 않으면 완성이다.

1. Combine the raspberries and sugar in a bowl and let them sit overnight in the refrigerator to release the juices.

   TIP. You can replace raspberries with berry mix or other fruits as you desire.

2. Drain the juice and bring to boil the juice. When it reduced add the raspberry pulp and return to a boil.

3. Drop a jam onto the chilled plate, if it formed a gel remove from the heat.

(7) 비건 마요네즈

Vegan mayonnaise

재료	**Ingredients**
두유 200g	Soy milk 200g
레몬즙 또는 식초 30g	Lemon juice or vinegar 30g
소금 6g	Salt 6g
아가베 시럽 20g	Agave Syrup 20g
씨겨자 30g	Whole grain mustard 30g
포도씨유 200g	Raisin seed oil 200g

Let's Make

1. 모든 재료를 핸드 블렌더에 넣고 갈아준다.

1. Blend all ingredients with a hand blender.

비건 치폴레 마요네즈 Vegan chipotle mayonnaise

재료

비건 마요네즈 200g

다진 치폴레(훈연 파프리카) 60g

Ingredients

Vegan mayonnaise 200g

Chopped chipotle pepper in
adobo sauce 60g

Let's Make

1. 만들어둔 비건 마요네즈에 곱게 다진 치폴레를 넣고 고르게 섞는다.

1. Put all ingredients in a food processor until combined.

---

(8) 비건 파마산 치즈

Vegan parmesan

재료

아몬드가루 130g

뉴트리션이스트(영양 효모) 30g

소금 8g

양파가루 5g

설탕 5g

Ingredients

Almond powder 130g

Nutritional yeast powder 30g

Salt 8g

Onion powder 5g

Sugar 5g

Let's Make

1. 블렌더에 모든 재료를 넣고 잘 섞일 때까지 갈아준다.

2. 밀폐 용기에 담아 냉장고에 보관한다.

1. Blend all ingredients with a hand blender.

2. Transfer into an airtight container and refrigerate it.

(9) 비건 소보로

Vegan crumble

재료	Ingredients
비건 버터 100g	Vegan butter 100g
땅콩버터 65g	Peanut butter 65g
소금 3g	Salt 3g
비정제 설탕 100g	Unrefined sugar 100g
달걀물 대체재 30g	Vegan egg 30g
(두유 27g+아마씨가루 3g)	(Ground flaxseed 3g+soy milk 27g mixture)
옥수수가루 95g	Cornmeal 95g
박력분 230g	Cake flour(Soft flour) 230g
베이킹파우더 6g	Baking powder 6g
베이킹소다 3g	Baking soda 3g

Let's Make

1. 옥수수가루, 박력분, 베이킹파우더, 베이킹소다를 체로 친다.
2. 가정용 믹싱기 볼에 비건 버터와 땅콩버터를 넣고 나뭇잎 모양의 비타를 이용해 부드럽게 풀어준다.
3. 2번 볼에 비정제 설탕과 소금을 넣어 다시 휘핑한 후 마요네즈처럼 부드러운 상태가 되면 달걀물 대체재를 넣어 섞는다.
4. 체로 친 1번의 가루를 넣고 비타로 섞어 보슬보슬한 소보로를 만든다.
5. 사용 후 남은 비건 소보로는 냉동실에 넣어 보관하고 사용 전 해동하면 된다.

1. Sift the cornmeal, cake flour, baking powder, baking soda.
2. In a mixing bowl, beat the vegan butter and peanut butter together.
3. Add the sugar and salt, beat until smooth like the mayonnaise texture. Add the vegan egg beat until evenly combined.
4. Over the same bowl, pour the 1 and mix till the mixture is crumbly.
5. Freeze leftover crumble, defrost before use. You can reuse the crumble.

비건 베이킹 시작 전 꼭 알아야 할 주의 사항

비건 베이킹 시작 전 꼭 알아야 할 주의 사항

A

반죽의 최종 온도는 26℃가 넘지 않게 주의한다. 반죽의 온도에 따라 1차 발효 시간이 변동될 수 있다. 반죽 온도에 따라 이스트의 발효력이 달라지므로 상황에 맞게 조정해주어야 반죽마다 일정한 발효 상태를 맞출 수 있다. 반죽 최종 온도가 높으면 1차 발효 시간을 다소 줄이고 서늘한 곳에서 발효한다. 반대로, 반죽 온도가 낮으면 1차 발효 시간을 늘리고 따뜻한 곳에서 보관해야 한다. 가정에서는 반죽 최종 온도가 희망 온도보다 높은 경우가 대부분이므로 반죽 시 여름에는 얼음물을, 겨울에는 차가운 물을 사용하는 것을 추천한다.

Be careful that the final temperature of the does not exceed 26℃/79℉. The First fermentation(bulk) time may vary depending on the temperature of the dough. Since the fermentation degree by yeast varies depending on the temperature of the dough, it must be adjusted according to the situation to achieve the constancy of fermentation.
If the final temperature of the dough is higher than desired temperature, reduce the fermentation time and store the dough in a cool place. On the contrary, if it is low, increase the fermentation time and store it in a warm place. In most cases, final temperature is higher than the desired temperature to home baker, so it is recommended to use ice water in summer and cold water in winter for mixing.

B

믹싱 시 사용하는 장비에 따라 믹싱 속도와 믹싱 시간이 달라질 수 있다. 스파 스탠드 믹싱기를 사용하는 경우, 레시피에 제시한 대로 1단, 2단으로(저속·중속·고속으로 변경) 사용하면 된다. 키친에이드, 캔우드, 위즈웰 등의 가정용 믹싱기를 사용할 수 있는데, 최고 믹싱 속도는 8~10단으로 다양하다. 그런 경우 1단은 2단 정도로, 2단은 4~5단계 정도로 조절해서 사용한다.

Depends on your baking equipment the mixing speed may variate. If you use spar stand mixer, you may follow the process as suggested. You can use variety of mixing devices, such as Kitchenaid, Kenwood, Wiswell for example he mixing speed may up to 8 or 10. You can take 2 as 1 and 4 or 5 to 2.

C

믹싱 시 믹싱 볼에 **액체 재료를 먼저 넣고** 가루 재료를 넣는 것이 좋다. 가루 재료 먼저 넣으면 훅이 움직이면서 물과 재료가 믹싱 볼 벽 위쪽까지 묻어 스크래퍼로 긁어줘야 하는 번거로움이 있다. 되도록 액체 재료부터 믹싱 볼에 넣도록 한다.

When mixing, it is recommended to add liquid ingredients first. If you put the dry ingredients first, while the mixer is working the mixture reaches high side of bowl which is inconvenient process to clean the side. It is preferred to put liquid ingredients first

D

대부분의 경우, **반죽에 넣는 충전물이나 두유액은 사용 전까지 냉장고에서 차갑게** 보관한다. 반죽 시 믹싱기를 사용하면 마찰열이 발생해 반죽 온도가 희망 온도보다 높아지는 경우가 대다수다. 이때 충분히 차가운 상태의 충전물을 사용하면 반죽의 최종 온도를 낮추는 데 도움이 된다. 반면 겨울철이나 온도가 아주 낮은 주방에서 빵을 만들 경우, 반죽 온도가 희망 온도보다 낮아지게 된다. 이때는 반죽에 넣기 전 충전물의 온도를 높여 반죽 온도를 높일 수 있다.

Filling or liquid ingredients into dough should be stored in the refrigerator until use.When using a kneading machine during mixing, it is common for the Final dough temperature to be higher than the desired temperature due to friction heat. In this case, using refrigerated ingredients will help lower the final temperature of the dough. On the other hand, when bread is made in winter or in well air-conditioned workplace, the dough temperature becomes lower than the desired temperature. In this case, the dough temperature can be increased by heating the ingredients before putting them into the dough.

건과일이나 곡류, 씨앗을 충전물로 사용할 경우 반죽 하루 전날 물에 불린다. 건조한 성분의 충전재가 반죽 속에 섞이면서 반죽의 수분을 빼앗는 경우가 많은데, 이를 방지하기 위함이다.

If dried fruits, cereals or seeds are used as filling ingredients, pre-soak them in water the day before use; to prevent dry ingredients from taking away moisture from the dough.

견과류를 이용하는 빵을 만들 경우 필요할 때마다 조금씩 견과류를 굽는 것이 번거로울 수 있으니, **한번에 많은 양을 구워두었다 꺼내 사용하면 편리**하다. 여름에는 견과류가 산폐되기 쉬우므로 보관에 주의해야 한다. 냉동실에 보관하는 편이 좋다.

If you use nuts for the recipes, it's cumbersome to bake as much as you need. So, prepare a lot at once. In summer, nuts turn rancid easily, so it's better to keep them in the freezer.

G

시중 크랜베리는 보통 당절임이 되어 있는데, 이때 사용한 설탕에 탄화 골분(동물의 뼈를 태워서 생성된 가루)이 들어 있을 수도 있다. 당절임 시 사용한 설탕의 성분을 정확히 알 수 없는 경우, 건무화과로 대체하는 것을 추천한다.

Commercially available cranberries are usually sweetened, and the sugar used during the process may contain carbonized bone powder. (Cane sugar is processed using animal bones-bone char). If you are concerned about used sugar, it is recommended to replace it with dried fig.

발효실이 없는 경우 간이 발효 박스(P.012)를 만들어 온습도계를 넣고 뜨거운 물을 갈면서 온습도를 맞춰준다.

If there is no dough proofer/fermentation chamber, make a simple homemade proofing box with a thermo-hygrometer and adjust the temperature and humidity with changing hot water.

2차 발효가 10분 남았을 때 미리 오븐을 예열한다. 오븐을 열 때, 온도가 20~30°C 정도 떨어진다. 원하는 온도보다 20~30°C 높게 예열한 후 반죽을 넣고 원하는 온도로 내려 굽는 것이 좋다. 단, 하드 빵을 오븐으로 구울 경우 오븐 속에 돌판과 자갈을 넣게 되므로 250°C 이상에서 적어도 1시간은 예열한다. 돌판이 충분히 달궈지고 자갈도 열을 충분히 머금어야 하기에 최소 50분 동안은 예열이 필요하다.

Preheat the oven in advance when the baking is about 10 minutes away. When you open the oven, there is a heat loss of about 20 to 30°C. Therefore, preheat the oven about 20 to 30°C above the temperature you want to bake. And put the dough in and lower the thermostat to the desired temperature. To make a crusty bread, add a baking stone and pebbles to the bottom rack of your oven and preheat the oven to above the 480°F at least 1 hour. This sufficiently heated stones will take at least 50 minutes to heat and help maintain the temperature of the oven.

제빵은 같은 레시피여도 그날의 온도, 습도, 오븐의 성능에 따라 결과물에 큰 차이가 날 수 있으니 실패했다고 포기하지 말고 꾸준히 도전해보세요.

Essential things before you start vegan baking

비건 베이킹 시작 전 꼭 알아야 할 주의 사항

PART 1

VAKE VEGAN BAKING

매일 먹어도 부담 없는
데일리 빵

플레인 식빵 | 녹차 크랜베리 식빵 | 가나슈 식빵 |

단호박 식빵 | 올리브 베이글 | 호두 크랜베리 베이글 |

시금치 치아바타 | 올리브 치아바타 |

호두 크랜베리 깜빠뉴

VEGAN PLAIN BREAD

플레인 식빵

플레인 식빵은 밀가루, 두유, 비정제 설탕, 비건 버터 등의 재료를 최적의 비율로 넣고 만들어 편하게 매일 먹을 수 있는 빵이에요. 충전물이 없어 심심한 맛이지만 그만큼 부담 없이 뜯어 먹을 수 있다는 게 플레인 식빵의 매력 아니겠어요. 남은 빵은 오븐이나 팬에 노릇하게 구워 먹어도 좋고, 일정한 크기로 잘라서 샌드위치를 만들어도 좋답니다.

Plain bread is made of ingredients such as flour, soy milk, unrefined sugar, and vegan butter at a perfect ratio. It's something that you can eat casually everyday. I enjoy bread with fillings, but I also like to eat slices of plain bread because of its simple nature. Leftover bread can be enjoyed again by heating it in an oven or on a frying pan, and you can also cut it into a certain size to make sandwiches.

Yield	250g 4개	250g * 4 loaves

Ingredient		
	반죽 재료	**Dough**
	강력분 500g	Strong flour 500g
	소금 10g	Salt 10g
	비정제 설탕 70g	Unrefined sugar 70g
	드라이이스트 7g	Dry yeast 7g
	조청 20g	Brown rice syrup 20g
	두유 175g	Soy milk 175g
	물 175g	Water 175g
	비건 버터 60g	Vegan butter 60g
	두유액	**Soy milk wash**
	두유 30g	Soy milk 30g
	아가베 시럽 15g	Agave syrup 15g

Pre-Check	**주의 사항 P.035**	**Notice P.035**
	Ⓐ + Ⓑ + Ⓒ +	Ⓐ + Ⓑ + Ⓒ +
	Ⓓ + Ⓗ + Ⓘ 필독!	Ⓓ + Ⓗ + Ⓘ read the suggestion!

매일 먹어도 부담 없는 데일리 빵

Recipe
Timeline

한눈에 보는 레시피 타임라인

① **반죽하기**
Mixing

② **둥글리기**
Rounding

③ **1차 발효하기(60min)**
Primary Fermentation/Bulk(60min)

④ **가스 빼기**
De-gas

⑤ **분할하기**
Dividing

⑥ **둥글리기**
Pre-shaping

⑦ **휴지하기(20min)**
Resting(20min)

⑧ **성형하기**
Shaping and Panning

⑨ **2차 발효하기(90min)**
Final Fermentation/Proofing(90min)

⑩ **굽기(170℃, 25min)**
Baking(340°F/170℃, 25min)

⑪ **식힘망에서 식히기**
Cooling

1   믹싱 볼에 물과 두유를 먼저 넣고 비건 버터를 제외한 반죽 재료를 넣어 저속으로 3분 정도 믹싱한다. 이어서 중속으로 3분 정도 더 믹싱한다.

2   비건 버터를 넣고 중속으로 6~7분 더 믹싱한다.

3   반죽 표면에 윤이 나고 사진과 같은 얇은 글루텐이 생기면 믹싱기를 멈춘다.

4   완성된 반죽은 가장자리를 중심으로 모아 둥근 모양으로 만든 후 몸 쪽으로 당겨 단단하게 둥글려준다.

1   Pour the water and soy milk into a mixing bowl then add the remaining dough ingredients except vegan butter. Knead for 3 minutes on low speed then 3 minutes on medium speed. Run the mixer on low speed to combine all ingredients evenly then develop the gluten on medium speed.

2   Add vegan butter and run for 6~7 minutes on medium speed.

3   If your dough is smooth and glossy and the gluten strands have developed such like photo, stop the mixer.

4   Turn dough onto a floured surface. Fold the sides in and over repeatedly then turn it around then pull the dough towards you. Repeat it several times until the gluten start to tighten.

5   둥글리기 한 반죽은 밑이 좁은 발효통에 넣고 실온에서 60분간 1차 발효한다.

       TIP. 소량인 경우 반죽을 둥글게 모아준 후 발효를 하면 퍼지지 않고 발효도 더 잘된다.

6   60분 발효 후 모습이다.

       TIP. 투명한 용기를 쓴 경우 아래쪽을 보면 작은 기포를 확인할 수 있다. 반죽의 부피가 1.5배 정도
       되면 완성된 것이다. 주방 온도가 낮다면 발효 시간이 늘어날 수 있다.

7   1차 발효가 끝나면 작업대와 반죽에 밀가루를 뿌리고 스크래퍼를 이용해 통에서 반죽
       을 꺼내 손바닥으로 평평하게 두드린다.

       TIP. 분할하기 쉽도록 일정한 두께로 만든다.

8   반죽을 약 250g씩 4개로 분할한다.

---

5   Transfer the dough to a dough-rising bucket with a narrow bottom. Cover the bucket, and allow the dough to rise about 1 hours.

       TIP. If your made small amount, shape the dough into a ball for Primary Fermentation. The dough will rise up instead of sideways and rise correctly.

6   After 1 hours fermentation.

       TIP. If you use a transparent container, you may observe the small bubbles at the bottom. it is completed when the dough rise about 1.5 times in bulk. It may rise even more slowly in a cool kitchen.

7   At the end of the rise, lightly flour the dough, turn the dough out onto a lightly floured surface using plastic scraper.

       TIP. If you make it even thickness, it is easy to divide by the same weight.

8   Divide it into four of 250g each and shape into round and firm balls.

Let's Bake 2

9   단단하게 둥글기 한 후 20분간 휴지한다.

10   휴지 후 반죽에 밀가루를 가볍게 뿌리고 이음매 부분이 위쪽으로 오게 작업대에 얹은 다음 밀대를 이용해 균일한 두께로 밀며 기포를 빼준다.

11   반죽을 세로로 길게 놓고 윗부분을 약간 넓게 잡아 살짝 당긴 후 둥근 반죽을 사각형이 되게 한다.

9   Rest the pre-shaped dough about 20 minutes.

10   Transfer the dough to a lightly floured surface. Using a dough roller to flatten and de-gas the dough into a long rectangle.

11   Making round edge to square, fold the dough towards the center.

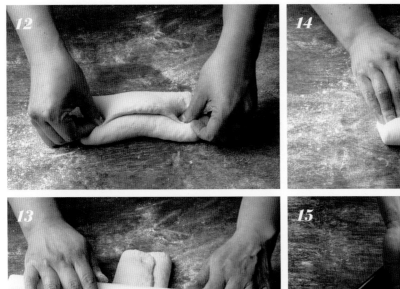

12   2/3 지점까지 접고 아래쪽도 동일하게 접어 3절접기 한다.

13   3절접기 한 반죽을 90도 돌려서 밀대로 가볍게 밀어준다.

14   반죽의 윗부분부터 풀리지 않게 잡아주면서 손끝으로 말아 내린다. 말아준 반죽은 이음매 부분이 풀리지 않게 손바닥이나 손가락으로 눌러 접착한다. 반죽을 굴려 이음매가 아래로 오도록 둔다.

15   2차 발효를 하기 위해 반죽을 120 × 110 × 80mm 크기의 메론식빵팬에 넣는다. 2차 발효가 되면서 틀의 모서리 부분까지 반죽이 잘 찰 수 있도록 균일한 힘을 가해 반죽 윗면을 가볍게 눌러준다.

12   Then fold the other edge of the dough over the folded portion. This is what we call single fold.

13   Place the dough vertically then flatten the dough with a dough roller.

14   Roll it down with your fingertips while holding the dough steady. Seal the seam pressing it firmly with the heel of your hand or fingertips. Turn the log over so its seam is on the bottom.

15   Place the log in a 5"×4"×3" loaf pan. (I used korean melon bread pan). Give a light pressure top of the dough to make sure the edges of the pan will fill up and rise evenly during the final fermentation.

Let's Bake 3

16 2차 발효는 온도 27~28℃, 습도 70~80%로 90분간 진행한다. 2차 발효 완료 10분 전에 컨벡션 오븐을 190℃로 예열한다.

17 2차 발효가 끝난 반죽은 성형 직후보다 2.5배 정도 부풀고, 가스가 차 가벼운 느낌이 다. 구움색이 잘 나도록 윗면에 붓으로 두유액을 발라준다.

18 오븐에 반죽을 넣고 170℃로 낮추어서 25분간 굽는다. 다 구워진 식빵은 틀째로 바닥 에 탁 쳐서 틀에서 분리한 다음 식힘망 위에서 식힌다.

16 Allow the dough to rise for 1 1/2 hours, in a condition of 80~82°F, 70~80% to humidity. When there's 10 minutes left to proof, preheat your oven to 375°F with a rack in the center.

17 At the end of proof, the dough risen about 2.5 times and feels soft, much lighter than when proof started. Brush the soy milk wash all over the top crust to lends it a nice golden color.

18 Reduce the oven temperature to 340°F and bake the bread for 25 minutes, until it's golden brown. Hit the bottom of pan after you take it out of the oven. Turn the bread out of the pan onto a rack to cool completely.

VEGAN MATCHA CRANBERRY BREAD

녹차 크랜베리 식빵

색이 고운 초록색 녹차 반죽에 빨간 크랜베리가 콕콕 박혀 있는 녹차 크랜베리 식빵은 알록달록한 색감만으로도 눈길을 사로잡아요. 쌉싸름한 녹차와 새콤달콤한 크랜베리의 조화를 느껴보세요.

Vividly red cranberries embedded within a gorgeous green coloured green tea dough is eye-catching for anyone. Enjoy the combination of the slightly bitter green tea and the sweet and sour cranberries.

Yield	약 300g 4개	300g * 4 loaves

Ingredient	**반죽 재료**	**Dough**
	강력분 500g	Strong flour 500g
	소금 10g	Salt 10g
	비정제 설탕 80g	Unrefined sugar 80g
	드라이이스트 8g	Dry yeast 8g
	녹차가루 12g	Matcha powder 12g
	조청 20g	Brown rice syrup 20g
	두유 175g	Soy milk 175g
	물 175g	Water 175g
	비건 버터 60g	Vegan butter 60g
	충전물	**Cranberry maceration**
	크랜베리 150g	Dried cranberry 150g
	물 40g	Water 40g
	두유액	**Soy milk wash**
	두유 30g	Soy milk 30g
	아가베 시럽 15g	Agave syrup 15g

Pre-Check	**주의 사항 P.035**	**Notice P.035**
	Ⓐ + Ⓑ + Ⓒ + Ⓓ + Ⓔ	Ⓐ + Ⓑ + Ⓒ + Ⓓ + Ⓔ
	Ⓖ + Ⓗ + Ⓘ 필독!	Ⓖ + Ⓗ + Ⓘ read the suggestion!

Recipe Timeline

한눈에 보는 레시피 타임라인

① **반죽하기**
Mixing

② **둥글리기**
Rounding

③ **1차 발효하기(60min)**
Primary Fermentation/Bulk(60min)

④ **가스 빼기**
De-gas

⑤ **분할하기**
Dividing

⑥ **둥글리기**
Pre-shaping

⑦ **휴지하기(20min)**
Resting(20min)

⑧ **성형하기**
Shaping and Panning

⑨ **2차 발효하기(90min)**
Final Fermentation/Proofing(90min)

⑩ **굽기(170℃, 25min)**
Baking(340°F/170℃, 25min)

⑪ **식힘망에서 식히기**
Cooling

Pre-Cook

미리 준비할 것

1. **충전물** : 물이 크랜베리에 충분히 흡수될 수 있도록 전날 미리 크랜베리에 물을 붓고 잘 섞어 사용 전까지 냉장고에 차갑게 보관한다(최소 3시간 전에 작업해두면 좋다).

1. **Cranberry maceration** : Mix all ingredients and leave to macerate for few hours. ⇨ Refrigerate the mixture before use

Let's Bake 1

1 믹싱 볼에 물과 두유를 먼저 넣고 비건 버터를 제외한 나머지 반죽 재료를 넣는다. 재료가 균일하게 잘 섞일 수 있도록 저속으로 3분 믹싱하고, 중속으로 3분 더 믹싱해 글루텐을 형성시킨다.

2 비건 버터를 넣고 중속으로 6~7분 더 믹싱한다. 반죽 표면에 윤이 나고, 얇은 글루텐 (P.042)이 생기면 믹싱기를 멈춘다.

3 미리 준비한 크랜베리 충전물을 넣고 반죽에 고루 섞일 때까지 저속으로 믹싱한다.

1 Pour the water and soy milk into a mixing bowl then add the remaining dough ingredients except vegan butter. Run the mixer on low speed for 3 minutes to combine all ingredients evenly then develop the gluten on medium speed for 3 minutes.

2 Add vegan butter and run for 6~7 minutes on medium speed. If your dough is smooth and glossy and the gluten strands have developed such like photo, stop the mixer.

3 Add the cranberry maceration and run the mixer on low speed until evenly combined.

Let's Bake 2

4 완성된 반죽은 가장자리를 중심으로 모아 둥근 모양으로 만든 후 몸 쪽으로 당겨 단단하게 둥글려준다.

5 둥글리기 한 반죽은 밑이 좁은 발효통에 넣고 실온에서 60분간 1차 발효한다.

TIP. 소량인 경우 반죽을 둥글게 모아준 후 발효를 하면 퍼지지 않고 발효도 더 잘된다.

6 60분 발효 후 모습이다.

TIP. 투명한 용기를 쓴 경우 아래쪽을 보면 작은 기포를 확인할 수 있다. 반죽의 부피가 1.5배 정도가 되면 완성된 것이다. 주방 온도가 낮다면 발효 시간이 늘어날 수 있다.

4 Turn dough onto a floured surface. Fold the sides in and over repeatedly then turn it around then pull the dough towards you. Repeat it several times until the gluten start to tighten.

5 Transfer the dough to a dough-rising bucket with a narrow bottom. Cover the bucket, and allow the dough to rise about 1 hours.

TIP. If your made small amount, shape the dough into a ball for Primary Fermentation. The dough will rise up instead of sideways and rise correctly.

6 After 1 hours fermentation.

TIP. If you use a transparent container, you may observe the small bubbles at the bottom. it is completed when the dough rise about 1.5 times in bulk. It may rise even more slowly in a cool kitchen.

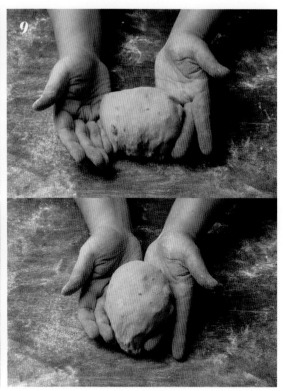

7   1차 발효가 끝나면 작업대와 반죽에 밀가루를 뿌리고 스크래퍼를 이용해 통에서 반죽
    을 꺼내 밀대로 민다.
    TIP. 분할하기 쉽도록 일정한 두께로 만든다.

8   반죽을 약 300g씩 4개로 분할한다.

9   단단하게 둥글리기 한 후 20분간 휴지한다.

7   At the end of the rise, lightly flour the dough, turn the dough out onto a lightly floured surface using plastic
    scraper.
    TIP. If you make it even thickness, it is easy to divide by the same weight.

8   Divide it into four of 300g each and shape into round and firm balls.

9   Rest the pre-shaped dough about 20 minutes.

Let's Bake 3

10 휴지 후 반죽에 밀가루를 가볍게 뿌리고 이음매 부분이 위쪽으로 오게 작업대에 얹은 다음 밀대를 이용해 균일한 두께로 밀며 기포를 빼준다.

11 반죽을 세로로 길게 놓고 윗부분을 약간 넓게 잡아 살짝 당긴 후 둥근 반죽을 사각형이 되게 한다.

12 2/3 지점까지 접고 아래쪽도 동일하게 접어 3절접기 한다

13 3절접기 한 반죽을 90도 돌려서 밀대로 가볍게 밀어준다. 반죽의 윗부분부터 풀리지 않게 잡아주면서 손끝으로 말아 내린다.

10 Transfer the dough to a lightly floured surface. Using a dough roller to flatten and de-gas the dough into a long rectangle.

11 Making round edge to square, fold the dough towards the center.

12 Then fold the other edge of the dough over the folded portion. This is what we call single fold.

13 Place the dough vertically then flatten the dough with a dough roller. Roll it down with your fingertips while holding the dough steady.

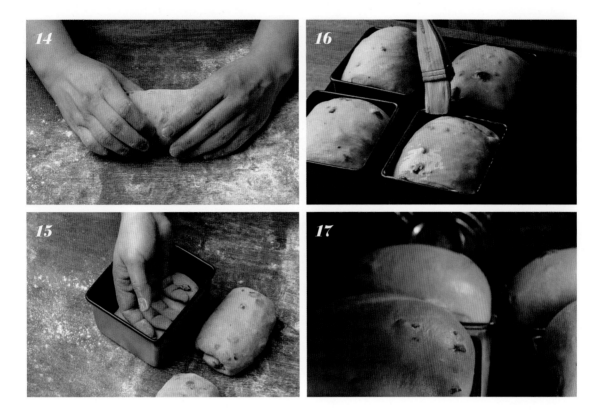

14 말아준 반죽은 이음매 부분이 풀리지 않게 손바닥이나 손가락으로 눌러 접착한다. 반죽을 굴려 이음매가 아래로 오도록 둔다.

15 2차 발효를 하기 위해 반죽을 120 × 110 × 80mm 크기의 메론식빵팬에 넣는다. 2차 발효가 되면서 틀 모서리 부분까지 반죽이 잘 차도록 윗면을 가볍게 눌러준다

16 2차 발효는 온도 27~28℃, 습도 70~80%로 90분간 진행한다. 2차 발효 완료 10분 전에 컨벡션 오븐을 190℃로 예열하고 발효가 끝나면 반죽 윗면에 두유액을 발라준다.
TIP. 2차 발효가 끝난 반죽은 성형 직후보다 2.5배 정도 부풀고, 가스가 차 가벼운 느낌이다.

17 오븐에 반죽을 넣고 170℃로 낮추어서 25분간 굽는다. 다 구워진 식빵은 틀째 바닥에 탁 쳐서 틀에서 분리한 다음 식힘망 위에서 식힌다.

14 Seal the seam pressing it firmly with the heel of your hand or fingertips. Turn the log over so its seam is on the bottom.

15 Place the log in a 5"×4"×3" loaf pan. (I used korean melon bread pan). Give a light pressure top of the dough to make sure the edges of the pan will fill up and rise evenly during the final fermentation.

16 Allow the dough to rise for 1 1/2 hours, in a condition of 80~82°F, 70~80% to humidity. When there's 10 minutes left to proof, preheat your oven to 375°F with a rack in the center. Brush the soy milk wash all over the top.
TIP. At the end of proof, the dough risen about 2.5 times and feels soft, much lighter than when proof started.

17 Reduce the oven temperature to 340°F and bake the bread for 25 minutes, until it's golden brown. Hit the bottom of pan after you take it out of the oven. Turn the bread out of the pan onto a rack to cool completely.

VEGAN GANACHE BREAD

가나슈 식빵

비건 식빵이 이런 맛이라고? 반죽에 코코아 파우더와 다크 초콜릿 칩만 넣는 것이 아니라, 굽고 난 식빵에 달콤 쌉싸름한 초코 크림을 듬뿍 넣어 마무리할 거예요. 초콜릿을 좋아하는 사람에게는 최고의 선물이 아닐까요. 달콤한 식빵 한 조각에 따뜻한 우유 한잔을 곁들이면 아이들 간식으로도 최고랍니다.

Vegan bread has never been better. It's not just cocoa powder and dark chocolate chips in the dough. The sweet, slightly bitter chocolate cream filled inside the baked bread will be a treat for anyone who enjoys chocolate. If you give your kids a glass of warm milk with a slice of this sweet, chocolate bread, they will reward you with chocolate smeared smiles sweeter than the bread itself.

Yield	265g 4개	265g * 4 loaves

Ingredient	**반죽 재료**	**Dough**
	강력분 500g	Strong flour 500g
	코코아가루 30g	Cacao powder 30g
	소금 10g	Salt 10g
	비정제 설탕 80g	Unrefined sugar 80g
	드라이이스트 8g	Dry yeast 8g
	두유 190g	Brown rice syrup 30g
	물 175g	Soy milk 190g
	조청 30g	Water 175g
	비건 버터 70g	Vegan butter 70g
	충전물	**Filling**
	다크 초콜릿 칩 150g	Dark chocolate chips 150g
	비건 초코 크림	**Vegan chocolate cream**
	만드는 법 P.028 참고	Recipe P.028
	두유액	**Soy milk wash**
	두유 30g	Soy milk 30g
	아가베 시럽 15g	Agave syrup 15g

Pre-Check	**주의 사항 P.035**	**Notice P.035**
	Ⓐ + Ⓑ + Ⓒ + Ⓓ +	Ⓐ + Ⓑ + Ⓒ + Ⓓ +
	Ⓗ + Ⓘ 필독!	Ⓗ + Ⓘ read the suggestion!

Recipe
Timeline

한눈에 보는 레시피 타임라인

① 반죽하기
Mixing

② 반죽 둥글리기
Rounding

③ 1차 발효하기(60min)
Primary Fermentation/Bulk(60min)

④ 가스 빼기
De-gas

⑤ 분할하기
Dividing

⑥ 둥글리기
Pre-shaping

⑦ 휴지하기(20min)
Resting(20min)

⑧ 성형하기
Shaping and Panning

⑨ 2차 발효하기(90min)
Final Fermentation/Proofing(90min)

⑩ 굽기(170℃, 25min)
Baking(340°F/170℃, 25min)

⑪ 식힘망에서 식히기
Cooling

⑫ 비건 초코크림 충전하기
Pipe the vegan chocolate cream

Let's Bake 1

1   믹싱 볼에 물과 두유를 먼저 넣고 비건 버터를 제외한 나머지 반죽 재료를 넣는다. 재료가 균일하게 섞일 수 있도록 저속으로 3분 믹싱하고, 중속으로 3분 더 믹싱해 글루텐을 형성시킨다.

2   비건 버터를 넣고 중속으로 6~7분 더 믹싱한다. 반죽 표면에 윤이 나고, 얇은 글루텐 (P.042)이 생기면 믹싱기를 멈춘다.

3   반죽에 다크 초콜릿 칩 충전물을 넣고 고르게 섞일 때까지 저속으로 믹싱한다.

4   완성된 반죽은 가장자리를 중심으로 모아 둥근 모양으로 만든 후 몸 쪽으로 당겨 단단하게 둥글려준다.

1   Pour the water and soy milk into a mixing bowl then add the remaining dough ingredients except vegan butter. Run the mixer on low speed for 3 minutes to combine all ingredients evenly then develop the gluten on medium speed for 3 minutes.

2   Add vegan butter and run for 6~7 minutes on medium speed. If your dough is smooth and glossy and the gluten strands have developed such like photo, stop the mixer.

3   Add the dark chocolate chips and run the mixer on low speed until evenly combined.

4   Turn dough onto a floured surface. Fold the sides in and over repeatedly then turn it around then pull the dough towards you. Repeat it several times until the gluten start to tighten.

5 둥글리기 한 반죽은 밑이 좁은 발효통에 넣고 실온에서 60분간 1차 발효한다.

TIP. 소량인 경우 반죽을 둥글게 모아준 후 발효를 하면 퍼지지 않고 발효도 더 잘된다.

6 60분 발효 후 모습이다.

TIP. 투명한 용기를 쓴 경우 아래쪽을 보면 작은 기포를 확인할 수 있다. 반죽의 부피가 1.5배 정도 되면 완성된 것이다. 주방 온도가 낮다면 발효 시간이 늘어날 수 있다.

7 1차 발효가 끝나면 작업대와 반죽에 밀가루를 뿌리고 스크래퍼를 이용해 통에서 반죽을 꺼내 손바닥으로 평평하게 두드린다.

TIP. 분할하기 쉽도록 일정한 두께로 만든다.

5 Transfer the dough to a dough-rising bucket with a narrow bottom. Cover the bucket, and allow the dough to rise about 1 hours.

TIP. If your made small amount, shape the dough into a ball for Primary Fermentation The dough will rise up instead of sideways and rise correctly.

6 After 1 hours fermentation.

TIP. If you use a transparent container, you may observe the small bubbles at the bottom. it is completed when the dough rise about 1.5 times in bulk. It may rise even more slowly in a cool kitchen.

7 At the end of the rise, lightly flour the dough, turn the dough out onto a lightly floured surface using plastic scraper.

TIP. If you make it even thickness, it is easy to divide by the same weight.

Let's Bake 2

8   반죽을 약 300g씩 4개로 분할한다.

9   단단하게 둥글리기 한 후 20분간 휴지한다.

8   Divide it into four of 300g each and shape into round and firm balls.

9   Rest the pre-shaped dough about 20 minutes.

10 　휴지 후 반죽에 밀가루를 가볍게 뿌리고 이음매 부분이 위쪽으로 오게 작업대에 얹은 다음 밀대를 이용해 균일한 두께로 밀며 기포를 빼준다.

11 　반죽을 세로로 길게 놓고 윗부분을 약간 넓게 잡아 살짝 당긴 후 둥근 반죽을 사각형이 되게 한다.

12 　2/3 지점까지 접고 아래쪽도 동일하게 접어 3절접기 한다.

13 　3절접기 한 반죽을 90도 돌려서 밀대로 가볍게 밀어준다. 윗부분부터 반죽이 풀리지 않게 잡아주면서 손끝으로 말아 내린다.

10 　Transfer the dough to a lightly floured surface. Using a dough roller to flatten and de-gas the dough into a long rectangle.

11 　Making round edge to square, fold the dough towards the center.

12 　Then fold the other edge of the dough over the folded portion. This is what we call single fold.

13 　Place the dough vertically then flatten the dough with a dough roller. Roll it down with your fingertips while holding the dough steady.

14 말아준 반죽은 이음매 부분이 풀리지 않게 손바닥이나 손가락으로 눌러 접착한다. 반죽을 굴려 이음매가 아래로 오도록 둔다.

15 2차 발효를 하기 위해 반죽을 120×110×80mm 크기의 메론식빵팬에 넣는다. 2차 발효되면서 틀 모서리 부분까지 반죽이 잘 차도록 윗면을 가볍게 눌러준다.

16 2차 발효는 온도 27~28℃, 습도 70~80%로 90분간 진행한다. 2차 발효 완료 10분 전 컨벡션 오븐을 190℃로 예열하고 발효가 끝나면 반죽 윗면에 두유액을 발라준다.
   TIP. 2차 발효가 끝난 반죽은 성형 직후보다 2.5배 정도 부풀고, 가스가 차 가벼운 느낌이다.

17 오븐에 반죽을 넣고 170℃로 낮추어서 25분간 굽는다. 다 구워진 식빵은 틀째 바닥에 탁 쳐서 틀에서 분리한 다음 완전히 식힌 후 초코 크림을 주입한다.

14 Seal the seam pressing it firmly with the heel of your hand or fingertips. Turn the log over so its seam is on the bottom.

15 Place the log in a 5"×4"×3" loaf pan. (I used korean melon bread pan). Give a light pressure top of the dough to make sure the edges of the pan will fill up and rise evenly during the final fermentation.

16 Allow the dough to rise for 1 1/2 hours, in a condition of 80~82°F, 70~80% to humidity. When there's 10 minutes left to proof, preheat your oven to 375°F with a rack in the center. Brush the soy milk wash all over the top.
   TIP. At the end of proof, the dough risen about 2.5 times and feels soft, much lighter than when proof started.

17 Reduce the oven temperature to 340°F and bake the bread for 25 minutes, until it's golden brown. Hit the bottom of pan after you take it out of the oven. Turn the bread out of the pan onto a rack to cool completely than pipe the chocolate cream.

VEGAN SUGAR PUMPKIN BREAD

단호박 식빵

맛있는 단호박을 발견했다면 식빵에도 넣어보세요. 초록색 호박씨와 노란 호박 퓨레를 넉넉히 넣어 호박의 풍미를 그대로 느낄 수 있어요. 슬라이스해 샌드위치 빵으로 활용해도 좋아요. 색감도 예쁘고 영양도 만점입니다. 남은 빵은 작게 잘라 오븐에 구워 수프에 뿌리는 크루통으로도 만들 수 있어요. 활용도 높은 단호박 식빵 만들어볼까요?

If you ever find ripe sweet pumpkins (specifically "Kabocha Squash"), mix it into the bread. Many people will enjoy the deep pumpkin flavor that comes from the green pumpkin seeds and the pumpkin puree. It's great when used for sandwich slices, and dicing and baking leftover slices can be used for crouton in your favorite soup. What's more is that it is not only delicious but also full of nutrition as well. This multi-purpose bread is worth your time in baking.

Yield	330g 4개	330g * 4 loaves

Ingredient

반죽 재료
강력분 500g
찐 단호박
(껍질 제외) 300g
소금 10g
비정제 설탕 80g
드라이이스트 8g
두유 180g
비건 버터 70g

충전물
구운 호박씨 150g
물 40g

두유액
두유 30g
아가베 시럽 15g

Dough
Strong flour 500g
Steamed kent pumpkin
(Japanese pumpkin) flesh 300g
Salt 10g
Unrefined sugar 80g
Dry yeast 8g
Soy milk 180g
Vegan butter 70g

Filling
Toasted pumpkin seeds 150g
Water 40g

Soy milk wash
Soy milk 30g
Agave syrup 15g

Pre-Check

주의 사항 P.035
Ⓐ + Ⓑ + Ⓒ + Ⓓ + Ⓔ +
Ⓕ + Ⓗ + Ⓘ 필독!

Notice P.035
Ⓐ + Ⓑ + Ⓒ + Ⓓ + Ⓔ +
Ⓕ + Ⓗ + Ⓘ read the suggestion!

Recipe
Timeline

한눈에 보는 레시피 타임라인

① **반죽하기**
Mixing

② **반죽 둥글리기**
Rounding

③ **1차 발효하기(60min)**
Primary Fermentation/Bulk(60min)

④ **가스 빼기**
De-gas

⑤ **분할하기**
Dividing

⑥ **둥글리기**
Pre-shaping

⑦ **휴지하기(20min)**
Resting(20min)

⑧ **성형하기**
Shaping and Panning

⑨ **2차 발효하기(90min)**
Final Fermentation/Proofing(90min)

⑩ **굽기(170℃, 30min)**
Baking(340°F/170℃, 30min)

⑪ **식힘망에서 식히기**
Cooling

Pre-Cook

미리 준비할 것

1. **충전물** : 빵 만들기 하루 전 150℃로 예열한 컨벡션 오븐에 10분간 호박씨를 굽는다. ⇨ 막 구워진 호박씨에 물을 부어 불린다. ⇨ 불린 호박씨는 사용 전까지 냉장고에 차갑게 보관한 다.⇨ 수분을 흡수한 호박씨를 사용하면 식빵의 촉촉함이 오래 유지된다.

1. **Pumpkin seeds maceration** : It takes a long time to absorb water so allow them to soak for a day. Toast pumpkin seeds on a baking pan at 300°F for 10 minutes to enhance the flavor. Place toasted pumkin seeds in a bowl and pour water to macerate. Using macerated pumpkin seeds helps to keep crumb moist and to delay staling. Refrigerate the maceration until use.

Let's Bake 1

1   믹싱 볼에 물과 두유를 먼저 넣고 비건 버터를 제외한 나머지 반죽 재료를 넣는다. 재료가 균일하게 섞일 수 있도록 저속으로 3분 믹싱하고, 중속으로 3분 더 믹싱해 글루텐을 형성시킨다.

2   비건 버터를 넣고 중속으로 6~7분 더 믹싱한다. 반죽 표면에 윤이 나고, 얇은 글루텐 (P.042)이 생기면 믹싱기를 멈춘다.

3   미리 준비한 호박씨 충전물을 넣고 반죽에 고루 섞일 때까지 저속으로 믹싱한다.

1   Pour the water and soy milk into a mixing bowl then add the remaining dough ingredients except vegan butter. Run the mixer on low speed for 3 minutes to combine all ingredients evenly then develop the gluten on medium speed for 3 minutes.

2   Add vegan butter and run for 6~7 minutes on medium speed. If your dough is smooth and glossy and the gluten strands have developed such like photo, stop the mixer.

3   Add the pumpkin seeds maceration and run the mixer on low speed until evenly combined.

Let's Bake 2

4   완성된 반죽은 가장자리를 중심으로 모아 둥근 모양으로 만든 후 몸 쪽으로 당겨 단단하게 둥글려준다.

5   둥글리기 한 반죽은 밑이 좁은 발효통에 넣고 실온에서 60분간 1차 발효한다.

     TIP. 소량인 경우 반죽을 둥글게 모아준 후 발효를 하면 퍼지지 않고 발효도 더 잘된다.

6   60분 발효 후 모습이다.

     TIP. 투명한 용기를 쓴 경우 아래쪽을 보면 작은 기포를 확인할 수 있다. 반죽의 부피가 1.5배 정도되면 완성된 것이다. 주방 온도가 낮다면 발효 시간이 늘어날 수 있다.

4   Turn dough onto a floured surface. Fold the sides in and over repeatedly then turn it around then pull the dough towards you. Repeat it several times until the gluten start to tighten.

5   Transfer the dough to a dough-rising bucket with a narrow bottom. Cover the bucket, and allow the dough to rise about 1 hours.

     TIP. If your made small amount, shape the dough into a ball for Primary Fermentation The dough will rise up instead of sideways and rise correctly.

6   After 1 hours fermentation.

     TIP. If you use a transparent container, you may observe the small bubbles at the bottom. it is completed when the dough rise about 1.5 times in bulk. It may rise even more slowly in a cool kitchen.

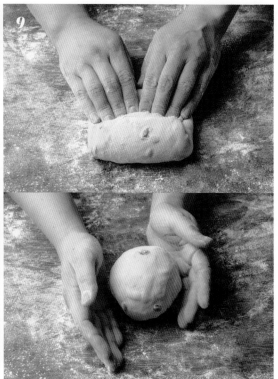

7 1차 발효가 끝나면 작업대와 반죽에 밀가루를 뿌리고 스크래퍼를 이용해 통에서 반죽을 꺼내 밀대로 민다.

TIP. 분할하기 쉽도록 일정한 두께로 만든다.

8 반죽을 약 330g씩 4개로 분할한다.

9 단단하게 둥글리기 한 후 20분간 휴지한다.

7 At the end of the rise, lightly flour the dough, turn the dough out onto a lightly floured surface using plastic scraper.

TIP. If you make it even thickness, it is easy to divide by the same weight.

8 Divide it into four of 330g each and shape into round and firm balls.

9 Rest the pre-shaped dough about 20 minutes.

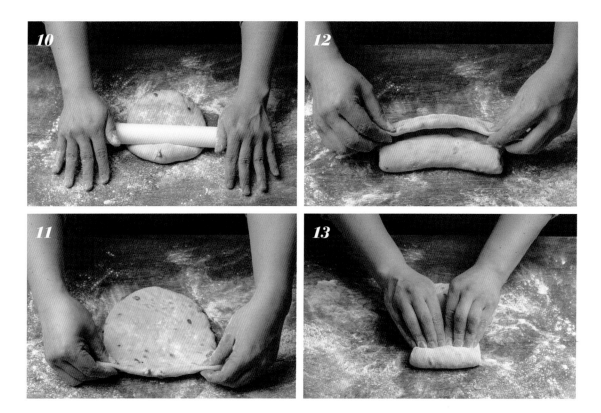

10 휴지 후 반죽에 밀가루를 가볍게 뿌리고 이음매 부분이 위쪽으로 오게 작업대에 얹은 다음 밀대를 이용해 균일한 두께로 밀며 기포를 빼준다.

11 반죽을 세로로 길게 놓고 윗부분을 약간 넓게 잡아 살짝 당긴 후 둥근 반죽을 사각형이 되게 한다.

12 2/3 지점까지 접고 아래쪽도 동일하게 접어 3절접기 한다.

13 3절접기 한 반죽을 90도 돌려서 밀대로 가볍게 밀어준다. 윗부분부터 반죽이 풀리지 않게 잡아주면서 손끝으로 말아 내린다.

10 Transfer the dough to a lightly floured surface. Using a dough roller to flatten and de-gas the dough into a long rectangle.

11 Making round edge to square, fold the dough towards the center.

12 Then fold the other edge of the dough over the folded portion. This is what we call single fold.

13 Place the dough vertically then flatten the dough with a dough roller. Roll it down with your fingertips while holding the dough steady.

14   말아준 반죽은 이음매 부분이 풀리지 않게 손바닥이나 손가락으로 눌러 접착한다. 반죽을 굴려 이음매가 아래로 오도록 둔다.

15   2차 발효를 하기 위해 반죽을 120 × 110 × 80mm 크기의 메론식빵팬에 넣는다. 2차 발효되면서 틀 모서리 부분까지 반죽이 잘 차도록 윗면을 가볍게 눌러준다.

16   2차 발효는 온도 27~28℃, 습도 70~80%로 90분간 진행한다. 2차 발효 완료 10분 전 컨벡션 오븐을 190℃로 예열하고 발효가 끝나면 반죽 윗면에 두유액을 발라준다.

      TIP. 2차 발효가 끝난 반죽은 성형 직후보다 2.5배 정도 부풀고, 가스가 차 가벼운 느낌이다.

17   오븐에 반죽을 넣고 170℃로 낮추어서 30분간 굽는다. 다 구워진 식빵은 틀째 바닥에 탁 쳐서 틀에서 분리한 다음 식힘망 위에서 식힌다.

14   Seal the seam pressing it firmly with the heel of your hand or fingertips. Turn the log over so its seam is on the bottom.

15   Place the log in a 5"×4"×3" loaf pan. (I used korean melon bread pan). Give a light pressure top of the dough to make sure the edges of the pan will fill up and rise evenly during the final fermentation.

16   Allow the dough to rise for 1 1/2 hours, in a condition of 80~82°F, 70~80% to humidity. When there's 10 minutes left to proof, preheat your oven to 375°F with a rack in the center. Brush the soy milk wash all over the top.

      TIP. At the end of proof, the dough risen about 2.5 times and feels soft, much lighter than when proof started.

17   Reduce the oven temperature to 340°F and bake the bread for 30 minutes, until it's golden brown. Hit the bottom of pan after you take it out of the oven. Turn the bread out of the pan onto a rack to cool completely.

VEGAN OLIVE BAGEL

올리브 베이글

제가 가장 아끼는 레시피예요. 올리브를 듬뿍 넣고 비건 파마산 치즈와 후추를 갈아 넣어, 은은한 후추 향이 매력적인 베이글이죠. 반죽에 올리브를 넣어 스프레드 없이 빵만 먹어도 기분 좋게 짭짤한 맛을 즐길 수 있어요. 특히 베이글 샌드위치를 만들 때, 이 올리브 베이글을 추천해요.

This recipe is one of my favorites. Plenty of olives, vegan parmesan cheese and ground black peppers make up a subtle pepper-y scent that is so enticing. You don't need a spread to enjoy this bagel, as the salt from the olive seasons this bagel to perfection. When making bagel sandwiches, this bagel is the way to go.

Yield	105g 10개	105g * 10 bagels

Ingredient	**반죽 재료**	**Dough**
	강력분 500g	Strong flour 500g
	소금 10g	Salt 10g
	비정제 설탕 30g	Unrefined sugar 30g
	비건 파마산 치즈 20g	Vegan parmesan cheese 20g
	후추 2g	Freshly ground black pepper 2g
	세미 드라이이스트 골드 3g	Semi-dry yeast gold label 3g
	물 310g	Water 310g
	비건 버터 40g	Vegan butter 40g
	충전물	**Filling**
	올리브 슬라이스 150g	Sliced olives 150g
	베이글 데칠 때 필요한 물	**Water bath**
	비정제 설탕 100g	Unrefined sugar 100g
	물 2ℓ	Water 2ℓ

Pre-Check	**주의 사항 P.035**	**Notice P.035**
	Ⓐ + Ⓑ + Ⓒ + Ⓓ +	Ⓐ + Ⓑ + Ⓒ + Ⓓ +
	Ⓗ + Ⓘ 필독!	Ⓗ + Ⓘ read the suggestion!

Recipe Timeline

한눈에 보는 레시피 타임라인

① **반죽하기**
Mixing

② **반죽 둥글리기**
Rounding

③ **1차 발효하기(30min)**
Primary Fermentation/Bulk(30min)

④ **가스 빼기**
De-gas

⑤ **분할하기**
Dividing

⑥ **예비 성형하기**
Pre-shaping

⑦ **휴지하기(20min)**
Resting(20min)

⑧ **성형하기**
Shaping and Panning

⑨ **2차 발효하기(30min)**
Final Fermentation/Proofing(30min)

⑩ **뜨거운 물에 데치기**
Water bath

⑪ **굽기(180℃, 13~15min)**
Baking(355°F/180℃, 13~15min)

⑫ **식힘망에서 식히기**
Cooling

Pre-Cook

미리 준비할 것

1. **충전물** : 올리브 슬라이스는 물기를 빼둔다.

2. **후추** : 통후추를 그라인더로 갈아서 준비한다. 후추분을 사용하면 텁텁한 맛이 나 풍미가 좋지 않다.

3. **종이 호일** : 종이 호일을 13×13cm 크기 정사각형으로 10개 잘라 준비한다.

1. **Filling(sliced olives)** : Drain the water from the can before using.

2. **Ground black pepper** : Freshly ground black pepper is a key to flavorful bagel. If you use pre-ground pepper, it gives dusty vibe on your mouth and bland flavor.

3. **Paper sheets** : Prepare 10 of 5×5 inch square parchment paper sheets.

Let's Bake 1

1   믹싱 볼에 물을 먼저 넣고 모든 반죽 재료를 넣는다. 재료가 균일하게 섞일 수 있도록 저속으로 3분 믹싱하고, 중속으로 8~10분 더 믹싱해 글루텐을 형성시킨다.
     TIP. 베이글 반죽은 수분율이 낮으므로 버터를 처음부터 넣고 믹싱한다.

2   반죽 표면에 윤이 나고, 사진과 같은 얇은 글루텐이 생기면 믹싱기를 멈춘다.
     TIP. 베이글 반죽은 수분이 적고 질감이 단단해 글루텐이 형성되기까지 시간이 오래 걸리는 편이다.

3   글루텐이 형성된 반죽을 작업대에 옮겨 넓게 펼친 후 그 위에 올리브를 고르게 뿌리고 반죽을 덮어 감싼다.

1   Pour the water into a mixing bowl then add the remaining dough ingredients. Knead for 3 minutes on low speed then 8 to 10 minutes on medium speed.
     TIP. The dough will be quite stiff; you can add butter from the beginning.

2   If your dough is smooth and glossy and the gluten strands have developed such like photo, stop the mixer.
     TIP. Bagel dough is low hydrated and are extremely stiff; it takes a bit more effort and time to develop the gluten.

3   Transfer the dough to the table then deflate it and sprinkle the olive slices over the dough then fold the dough to cover the olive.

4   스크래퍼로 반죽을 자르면서 손으로 직접 올리브를 섞는다.

       TIP. 올리브 충전물을 믹싱기를 이용해 섞지 않는 이유는 단단한 베이글 반죽과 수분이 많은 올리브가 믹싱 볼에서 만나 겉돌아서 재료가 잘 섞이지 않기 때문이다.

5   완성된 반죽은 둥글리기해 밑이 좁은 발효통에 넣고 실온에서 30분간 1차 발효한다.

       TIP. 수분이 날아가지 않도록 밀폐 용기에 넣거나 발효통에 넣어 꼼꼼히 랩핑한다.

6   30분 발효 후 모습이다.

       TIP. 30분 발효로는 외관상의 큰 변화를 확인할 수 없다.

4   Using a dough scraper cutting in the olives into the dough until well mixed.

       TIP. It isn't easy to mix wet ingredients evenly in a low hydration dough; Using a mixing machine to combine, the dough may be slip in the mixing bowl because of the moisture of the olive.

5   Gather the dough into a ball and place the dough to a dough-rising bucket. Cover the bucket, and allow the dough to rise about 30 minutes.

       TIP. It is recommended to use a dough container with lid or a small bowl.

6   After 30 minutes fermentation.

       TIP. There is no significant change in appearance after bulk.

7   1차 발효가 끝나면 작업대와 반죽에 밀가루를 뿌리고 스크래퍼를 이용해 통에서 반죽을 꺼내 손바닥으로 두드리거나 밀대로 민다.
TIP. 분할하기 쉽도록 일정한 두께로 만든다.

8   반죽을 105g씩 10개로 분할한다.

9   반죽을 타원형으로 예비 성형한 후 약 20분간 휴지한다.
TIP. 타원형으로 만드는 이유는 반죽을 길게 늘여서 양 끝을 이어 붙이면 우리가 아는 링 형태의 베이글이 나오기 때문이다.

10 휴지 후 반죽에 밀가루를 가볍게 뿌리고 이음매 부분이 위쪽으로 오게 작업대에 얹은 후 밀대를 이용해 균일한 두께로 밀며 기포를 빼준다.

7   At the end of the rise, turn the dough out onto a lightly floured surface using plastic scraper. Gently deflate the dough with your hands and make it even thickness.
TIP. If you make it even thickness, it is easy to divide by the same weight.

8   Divide it into ten of 105g each.

9   Shape into oval and rest the pre-shaped dough about 20 minutes.
TIP. Shape each dough into a rope form and seal the two ends together to make this transition easier, pre-shape the dough into oval.

10 Lightly flour the dough, transfer it to a clean work surface; turn upside down. De-gas the dough along the long side with a dough roller.

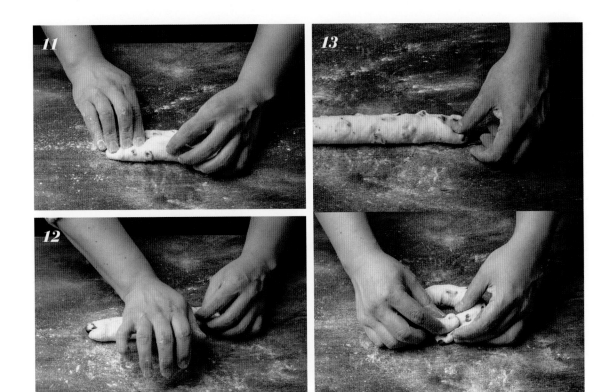

11 반죽을 가로로 작업대에 놓은 후 반죽 아래쪽을 2/3 지점까지 접고 반죽의 위쪽은 1/2 지점까지 내려 접는다.

12 반죽을 위에서 아래로 반 접은 후 이음매를 꾹꾹 눌러주며 길쭉하게 편다.

13 양 끝부분의 한쪽을 손으로 이어 붙여 지름 10cm의 링 형태로 성형한다.

11 Fold the down half up to the middle and seal and fold the top down to the middle again, slightly overlapping the first fold.

12 Fold the dough in half. Seal up entire length except the one end.

13 Fold each side of the dough then make the ends overlap. Seal the two ends together; shape a bagel about 4" in diameter.

14   링 형태를 손으로 꼼꼼히 다듬은 후 미리 준비한 종이 호일에 베이글의 이음매 부분이
아래로 향하도록 놓는다.

15   2차 발효는 온도 27~28℃, 습도 80% 상태에서 30분간 진행한다. 2차 발효 완료 10분
전 컨벡션 오븐을 200℃로 예열하고 베이글 데칠 물을 가열한다.

        TIP. 2차 발효가 끝난 반죽은 발효 전보다 부피가 커지고, 폭신하고 부드러운 느낌이다.

16   끓는 물에 종이 호일이 위로 가도록 반죽을 넣은 후 온도를 낮추기 위해 불을 끈다. 15
초간 데친 후 반죽을 뒤집고 다시 불을 켠 상태에서 15초간 더 데친다.

17   데친 반죽은 일정한 간격을 띄우고 팬닝한 후 오븐에 넣는다. 오븐 온도를 180℃로 낮추
어 노릇노릇한 색이 날 때까지 13~15분간 굽고 다 구워지면 식힘망으로 옮겨 식힌다.

14   Place the bagels onto each of the parchment paper sheets; seam side down.

15   Allow the dough to rise for 30 minutes, in a condition of 80~82°F, 80% to humidity. When there's 10 minutes
left to proof, preheat your oven to 390°F and prepare the water bath by heating the water, add the sugar.
Bring the mixture to a boil.

        TIP. At the end of proof, the dough is bulky and feels soft, much lighter than when proof started.

16   Transfer the bagels to the simmering water then turn off the heat to reduce the temperature. Cook the bagels
for 15 seconds, flip them over then turn on the heat and cook 15 seconds more.

17   Place them back on the baking pan leaving the spaces. Reduce the oven temperature to 355°F and bake the
bagels for 13 to 15 minutes, until they're lightly golden brown. Remove the bagels from the oven, and cool
completely on a rack.

VEGAN WALNUT CRANBERRY BAGEL

호두 크랜베리 베이글

새콤달콤한 크랜베리는 담백한 크림치즈와의 조합이 특히 좋고 다양한 재료를 곁들인 샌드위치와도 잘 어울려요. 크랜베리 대신 건포도, 건살구, 무화과 등 좋아하는 건과일과 견과류를 넣어 여러분만의 베이글을 만들어보는 것도 추천해요.

Sweet and sour cranberries go well with cream cheese, and also with sandwiches that have various ingredients. Make your own bagel with not only cranberries, but also raisins, dried apricots, figs, and other fruits and nuts as you like.

Yield	109g 10개	109g * 10 bagels

Ingredient	**반죽 재료**	**Dough**
	강력분 500g	Strong flour 500g
	소금 10g	Salt 10g
	비정제 설탕 30g	Unrefined sugar 30g
	드라이이스트 3g	Semi-dry yeast gold label 3g
	물 310g	Water 310g
	비건 버터 40g	Vegan butter 40g
	충전물	**Filling**
	건크랜베리 80g	Dried cranberry 80g
	구운 호두 80g	Toasted walnut pieces 80g
	물 45g	Water 45g
	베이글 데칠 때 필요한 물	**Water bath**
	비정제 설탕 100g	Unrefined sugar 100g
	물 2ℓ	Water 2ℓ

Pre-Check	**주의 사항 P.035**	**Notice P.035**
	Ⓐ + Ⓑ + Ⓒ + Ⓓ + Ⓔ +	Ⓐ + Ⓑ + Ⓒ + Ⓓ + Ⓔ +
	Ⓕ + Ⓖ + Ⓗ + Ⓘ 필독!	Ⓕ + Ⓖ + Ⓗ + Ⓘ read the suggestion!

Recipe Timeline

한눈에 보는 레시피 타임라인

① **반죽하기**
Mixing

② **반죽 둥글리기**
Rounding

③ **1차 발효하기(30min)**
Primary Fermentation/Bulk(30min)

④ **가스 빼기**
De-gas

⑤ **분할하기**
Dividing

⑥ **예비 성형 하기**
Pre-shaping

⑦ **휴지하기(20min)**
Resting(20min)

⑧ **성형하기**
Shaping and Panning

⑨ **2차 발효하기(30min)**
Final Fermentation/Proofing(30min)

⑩ **뜨거운 물에 반죽 데치기**
Water bath

⑪ **굽기(180℃, 13min)**
Baking(355°F/180℃, 13~15min)

⑫ **식힘망에서 식히기**
Cooling

Pre-Cook

미리 준비할 것

1. **충전물 :** 구운 호두와 크랜베리에 물을 넣고 잘 섞어준 후 3시간 이상 불린다. ⇨ 사용 전까지 냉장고에 차갑게 보관한다(사용 전날 미리 준비해두면 좋다).

2. **종이 호일 :** 종이 호일을 13×13cm 크기의 정사각형으로 10개 잘라 준비한다.

1. **Filling(walnut&cranberry) :** Mix cranberry, toasted walnut pieces and water; leave to macerate for few hours. Refrigerate the mixture before use.

2. **Paper sheets :** Prepare 10 of 5×5 inch square parchment paper sheets.

매일 먹어도 부담 없는 데일리 빵

Let's Bake 1

1   믹싱 볼에 물을 먼저 넣고 모든 반죽 재료를 넣는다. 재료가 균일하게 섞일 수 있도록 저속으로 3분 믹싱하고, 중속으로 8~10분 더 믹싱해 글루텐을 형성시킨다.
        TIP. 베이글 반죽은 수분율이 낮으므로 버터를 처음부터 넣고 믹싱한다.

2   반죽 표면에 윤이 나고, 얇은 글루텐(P.042)이 생기면 믹싱기를 멈춘다.
        TIP. 베이글 반죽은 수분이 적고 질감이 단단해 글루텐이 형성되기까지 시간이 오래 걸리는 편이다.

3   반죽이 충분히 쳐지면 전처리해둔 충전물을 넣고 저속으로 고르게 섞는다.

1   Pour the water into a mixing bowl then add the remaining dough ingredients. Knead for 3 minutes on low speed then 8 to 10 minutes on medium speed.
        TIP. The dough will be quite stiff; you can add butter from the beginning.

2   If your dough is smooth and glossy and the gluten strands have developed such like photo, stop the mixer.
        TIP. Bagel dough is low hydrated and are extremely stiff; it takes a bit more effort and time to develop the gluten.

3   Add the cranberry walnut maceration and run the mixer on low speed until evenly combined.

Let's Bake 2

4 완성된 반죽은 둥글리기 해 밑이 좁은 발효통에 넣고 실온에서 30분간 1차 발효한다.
  TIP. 수분이 날아가지 않도록 밀폐 용기에 넣거나 발효통에 넣어 꼼꼼히 랩핑한다.

5 30분 발효 후 모습이다.
  TIP. 30분 발효만으로는 외관상의 큰 변화를 확인할 수 없다.

6 1차 발효가 끝나면 작업대와 반죽에 밀가루를 뿌리고 스크래퍼를 이용해 통에서 반죽
  을 꺼내 손바닥으로 두드리거나 밀대로 민다.
  TIP. 분할하기 쉽도록 일정한 두께로 만든다.

4 Gather the dough into a ball and place the dough to a dough-rising bucket. Cover the bucket, and allow the dough to rise about 30 minutes.
  TIP. It is recommended to use a dough container with lid or a small bowl.

5 After 30 minutes fermentation.
  TIP. There is no significant change in appearance after bulk.

6 At the end of the rise, turn the dough out onto a lightly floured surface using plastic scraper. Gently deflate the dough with your hands and make it even thickness.
  TIP. If you make it even thickness, it is easy to divide by the same weight.

7   반죽을 110g씩 10개로 분할한다.

8   분할한 반죽은 타원형으로 예비 성형해 20분간 휴지한다.
     TIP. 타원형으로 만드는 이유는 반죽을 길게 늘여 양 끝을 이어 붙이면 우리가 아는 링 형태의 베이글
     이 나오기 때문이다.

9   휴지 후 반죽에 밀가루를 가볍게 뿌리고 이음매 부분이 위쪽으로 오게 작업대에 얹은
     다음 밀대를 이용해 균일한 두께로 밀며 기포를 빼준다.

10   반죽을 가로로 작업대에 놓고 반죽 아래쪽을 2/3 지점까지 접고 위쪽은 1/2 지점까지
     내려 접는다.

7   Divide it into ten of 110g each.

8   Shape into oval and rest the pre-shaped dough about 20 minutes.
     TIP. Shape each dough into a rope form and seal the two ends together; to make this transition easier, preshape the
     dough into oval.

9   Lightly flour the dough, transfer it to a clean work surface; turn upside down. De-gas the dough along the
     long side with a dough roller.

10   Fold the down half up to the middle and seal and fold the top down to the middle again, slightly overlapping
     the first fold.

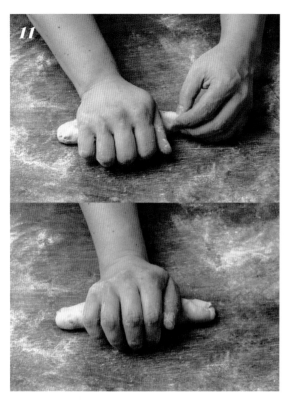

11   반죽을 위에서 아래로 반 접은 후 이음매를 꾹꾹 눌러주며 길쭉하게 편다.

12   양 끝부분의 한쪽은 손으로 눌러서 다른 쪽 반죽을 이어 붙일 수 있게 편 후 양 끝부분을 이어 붙여 지름 10cm의 링 형태로 성형한다.

11   Fold the dough in half. Seal up entire length except the one end.

12   Fold each side of the dough then make the ends overlap. Seal the two ends together; shape a bagel about 4" in diameter.

13 링 형태를 손으로 꼼꼼히 다듬은 후 미리 준비한 종이 호일에 베이글의 이음매 부분이 아래로 향하도록 놓는다.

14 2차 발효는 온도 27~28℃, 습도 80% 상태에서 30분간 진행한다. 2차 발효 완료 10분 전 컨벡션 오븐을 200℃로 예열하고 베이글 데칠 물을 가열한다.

   TIP. 2차 발효가 끝난 반죽은 발효 전보다 부피가 커지고, 폭신하고 부드러운 느낌이다.

15 끓는 물에 종이 호일이 위로 가도록 반죽을 넣은 후 온도를 낮추기 위해 불을 끈다. 15초간 데친 후 반죽을 뒤집고 다시 불을 켠 상태에서 15초간 더 데친다.

16 데친 반죽은 일정한 간격을 띄우고 팬닝한 후 오븐에 넣는다. 오븐 온도를 180℃로 낮추어 노릇노릇한 색이 날 때까지 13~15분간 굽고 다 구워지면 식힘망으로 옮겨 식힌다.

13 Place the bagels onto each of the parchment paper sheets; seam side down.

14 Allow the dough to rise for 30 minutes, in a condition of 80~82°F, 80% to humidity. When there's 10 minutes left to proof, preheat your oven to 390°F and prepare the water bath by heating the water, add the sugar. Bring the mixture to a boil.

   TIP. At the end of proof, the dough is bulky and feels soft, much lighter than when proof started.

15 Transfer the bagels to the simmering water then turn off the heat to reduce the temperature. Cook the bagels for 15 seconds, flip them over then turn on the heat and cook 15 seconds more.

16 Place them back on the baking pan leaving the spaces. Reduce the oven temperature to 355°F and bake the bagels for 13 to 15 minutes, until they're lightly golden brown. Remove the bagels from the oven, and cool completely on a rack.

VEGAN SPINACH CIABATTA

시금치 치아바타

치아바타는 하드빵에 속하지만 식감이 부드러워요. 밀가루, 소금, 이스트, 물, 올리브유만 있으면 부드러운 플레인 치아바타를 만들 수 있어요. 플레인 치아바타 레시피에 생시금치를 넣어 담백하게 먹을 수 있는 레시피를 소개할게요. 시금치를 팬에 살짝 볶아 소금과 후추로 간해 넣어주면 풍미가 한층 더 살아나요. 반죽할 때 시금치가루를 넣으면 색도 예쁘니 한번 시도해보세요.

Ciabatta is the softest bread among hard breads. You only need flour, salt, yeast, water, and olive oil to make soft and appetizing ciabattas. Plain ciabatta is good enough to be enjoyed on its own, but that is not why we are here. In order to make it more delectable, this book contains a spinach recipe that makes the simple ciabatta even better. Stir-fry spinach in a pan, season it with salt and pepper, and fold it while putting it in order to hold and trap the flavors inside. Try adding spinach powder when you knead the dough for a more gorgeous, green colored spinach ciabatta.

Yield	8개의 치아바타	8 ciabattas

Ingredient	**반죽 재료**	**Dough**
	트레디션 밀가루 200g	French bread flour (tradition française) 200g
	강력분 300g	High gluten flour(gruau flour) 300g
	비건 파마산 치즈 30g	Vegan parmesan cheese 30g
	물 340g	Water 340g
	소금 10g	Salt 10g
	드라이이스트 2g	Dry yeast 2g
	고생지 100g(생략가능)	Pâte fermentée 100g
	바시나주(★반죽 후반부에 넣는 물) 40g	Bassinage water 40g
	올리브유 50g	Olive oil 50g
	충전물	**Filling**
	생시금치 120g	Fresh spinach 120g

Pre-Check	**주의 사항 P.035**	**Notice P.035**
	Ⓐ + Ⓑ + Ⓒ + Ⓘ	Ⓐ + Ⓑ + Ⓒ + Ⓘ
	필독!	read the suggestion!

Recipe Timeline

한눈에 보는 레시피 타임라인

① **반죽하기**
Mixing

② **45분 후 사방접기**
Bulk(45min) Folding

③ **다시 45분 발효(1차 발효 총 90min)**
Bulk(45min)

④ **냉장고에 넣기 전 사방접기**
Folding

⑤ **다음 날까지 냉장고에서 1차 발효**
Bulk fermentation in the fridge
overnight

⑥ **분할하기, 팬닝**
Dividing and Panning

⑦ **2차 발효 1시간(온도 250℃로 오븐 예열하기)**
Pre-heat the oven (with baking stone)
during Proofing(60min)

⑧ **굽기(250℃, 오븐 끄고 5min)**
Baking(480˚F/250℃, 5min oven off

⑨ **2차 굽기(230℃, 8min)**
445˚F/230℃, 8min oven on)

⑩ **식힘망에서 식히기**
Cooling

Pre-Cook

미리 준비할 것

1. **충전물** : 시금치는 씻어서 물기를 뺀 후 잘라서 준비한다.
2. **사각 용기** : 치아바타는 별도의 성형 없이 네모난 형태로 잘라서 굽는 빵이므로 반죽을 담을
사각 용기를 사용하면 좋다.

1. **Filling** : Wash the spinach, remove excess water then chop roughly.
2. **Rectangular dough container** : There is no shaping for ciabatta, all you have to do
is divide the dough into squares. Using rectangular-types helps to form a square.

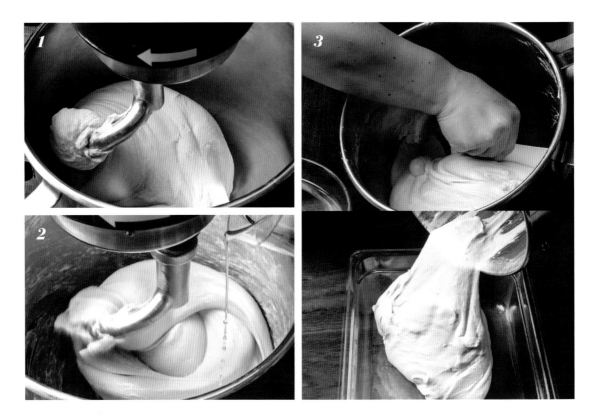

Let's Bake 1

1   믹싱 볼에 물을 먼저 넣고 바시나주, 올리브유를 제외한 나머지 반죽 재료를 넣는다. 재료가 균일하게 섞일 수 있도록 저속으로 3분 믹싱하고, 중속으로 6~7분 더 믹싱한다.

2   반죽 표면에 윤이 나고 얇은 글루텐이 생기면 믹싱기를 중속으로 맞추고 바시나주와 올리브유 순으로 소량씩 천천히 부어준다.

     TIP. 분량의 바시나주와 올리브유를 한번에 넣으면 반죽이 믹싱 볼에서 미끄러질 수 있다. 반죽에 물과 올리브유가 서서히 흡수될 수 있게 소량씩 부어주는 것이 중요하다.

3   1차 발효는 두 차례에 걸쳐 총 90분간 진행한다. 올리브유를 바른 사각 용기에 반죽을 붓고 사방접기해 반죽을 조심스럽게 뒤집은 후 45분간 1차 발효를 한다.

1   Pour the water into a mixing bowl then add the remaining dough ingredients except bassinage and olive oil. Knead for 3 minutes on low speed then about 6 to 7 minutes on medium speed.

2   If your dough gets smooth and glossy and the gluten strands have developed such like photo, add water and olive oil just a little bit at a time on medium speed.

     TIP. If you add lots of water or oil at once, dough spins in the bowl. It's important to pour in a small amount of water and olive oil into the dough so that it can absorbed slowly.

3   The first fermentation runs twice for 45 minutes, a total of 90 minutes. Transfer the dough to a rectangular dough container greased with olive oil. Then gently lift and fold dough over onto itself. Repeat it four sides. Then flip the dough over so the top is now the bottom. Let it rise for 45 minutes.

---

4   45분 후 잘라놓은 시금치를 반죽 위에 고르게 얹고 사방접기 한다. 접힌 부분이 풀리지 않도록 조심스럽게 반죽을 뒤집어 통에 넣은 후 다시 45분간 발효한다.

5   1차 발효 후 반죽을 다시 사방접기 해 표면에 올리브유를 바르고 비닐을 얹는다. 12~18시간 동안 냉장고(3℃)에서 천천히 발효할 수 있도록 한다.

     TIP. 냉장고가 가득 차 있다면 냉장 효율을 고려해 제시된 온도보다 냉장고 온도를 더 낮춘다. 꽉 찬 냉장고의 실제 내부 온도는 설정 온도보다 높을 수 있기 때문이다.

6   다음 날 냉장고에서 반죽을 꺼내 실온에 둔다. 여름철에는 30분, 겨울철에는 1시간가량 둔다.

---

4   After 45 minutes, sprinkle the chopped spinach over the dough. Then fold the four sides of dough and flip the dough over gently. Let it rise another 45 minutes.

5   After bulk fold the dough like just before then apply olive oil on the surface and cover the dough with plastic. Without lid, store in a 37℉ refrigerator overnight(12-18 hours) let it ferment slowly in the refrigerator.

     TIP. If the refrigerator is full, the actual internal temperature may be higher than the set point temperature. In that case, set the temperature of the refrigerator lower than the suggested temperature.

6   Remove the dough from the fridge, let the dough take the heat. Leave it for 30 minutes in summer and about 1 hour in winter.

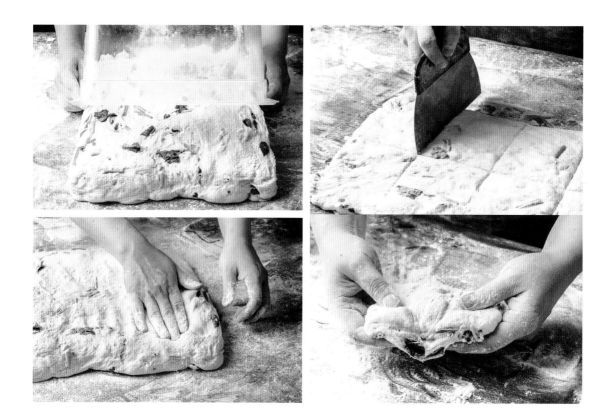

7 반죽과 작업대 위에 밀가루를 넉넉히 뿌리고 반죽이 들어 있는 통을 뒤집어 자연스럽게 반죽이 통에서 분리될 때까지 기다린다. 반죽이 통에서 완전히 분리되면 손에 들러붙지 않도록 반죽 위에 밀가루를 뿌린다.

8 두께가 균일해지도록 손바닥으로 반죽을 가볍게 두드리며 폭 26cm, 길이 40cm 크기의 직사각형으로 펴준다.

9 26cm 쪽은 2등분, 40cm 쪽은 4등분해 13×10cm 크기의 작은 직사각형 8개로 분할하고 반죽 윗면이 위로 향하도록 천 위에 놓아준다.

TIP. 천 위로 반죽을 옮길 때 분할한 모양이 변형되지 않도록 주의한다.

7 Uncover and sprinkle lots of flour on top of the dough, then turn it out onto a liberally floured work surface. Wait for the dough to fall out of the container intactly. When the dough is completely separated from the container, sprinkle flour on top so that the dough no longer sticks to the hand.

8 Gently deflate the dough with your hands into a 10"×16" rectangle and make it even thickness.

9 Divide the 10 inch side in half and the 16 inch side into four; divide into eight small rectangles, each 5" long × 4" wide. Laying them down on the couche the top of the dough faces upward.

TIP. Handling the dough gently, transfer each piece intactly on the couche.

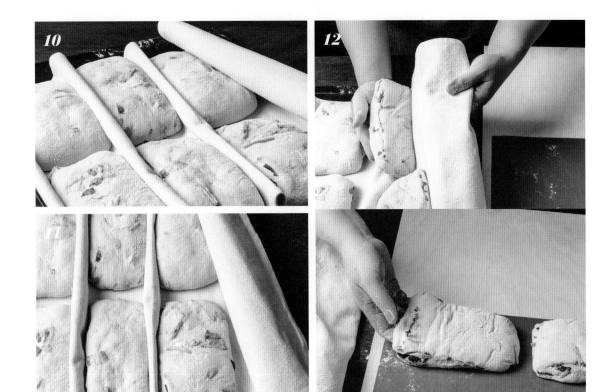

10   반죽 표면이 마르지 않도록 실온에서 1시간 동안 2차 발효한다.

         TIP. 실내가 건조하다면 반죽 위에 천을 덮어 마르지 않도록 한다.

11   2차 발효가 시작되면 오븐에 돌판과 자갈을 넣고 250℃로 예열하고, 발효가 끝나기 5분 전에는 주둥이가 긴 주전자에 물을 끓인다.

         TIP. 가지고 있는 오븐의 사양이 각기 다르므로 최고온도로 예열하면 된다.

12   2차 발효가 끝나면 반죽 아랫면이 위로 향하도록 베이킹 시트에 일정한 간격으로 놓고 반죽에 밀가루를 가볍게 뿌린다.

         TIP. 베이킹용 철판이나 넓은 나무판 위에 테프론 시트나 베이킹 시트를 놓고 반죽을 올리면 오븐에 밀어 넣기 편하다.

10   Keeping the surface moist then let the dough to rise for 1 hours at room temperature.

         TIP. To prevent the top forms a skin, cover loosely with a floured couche.

11   Towards the start of the rising time, place a baking stone in the middle of the oven then place a pan filled with pebbles below the stone preheat the oven to 480°F. Boil water in a long-mouth kettle about 5 minutes before fermentation is completed.

         TIP. Since the specifications of the oven are all different, preheat to the highest temperature of each oven.

12   At the end of proof, place top of the dough facing up on a baking sheet, leaving a space between them. You may lightly flour on a dough.

         TIP. Using a Teflon or baking sheet on a baking pan or wooden plate for proof, make it easier to push the dough into the oven.

Let's Bake 3

13　반죽이 놓인 베이킹 시트를 돌판 위에 밀어 넣는다.

14　주둥이가 긴 주전자에 담긴 뜨거운 물을 예열된 자갈 위에 붓고 재빨리 문을 닫는다.
TIP. 자갈 위에 올라오는 수증기로 화상을 입을 수 있으므로 반드시 오븐용 장갑을 끼고 작업한다.

15　컨벡션 오븐을 사용할 경우 바람에 의해 반죽 표면이 빠르게 마를 수 있어 반죽을 넣자마자 5분간 오븐을 꺼준다. 오븐을 꺼도 충분히 달궈진 돌판이 있어 반죽이 익기 시작한다.

16　5분 후 다시 오븐을 켜고 230℃로 맞춘 후 7~8분 더 굽는다. 다 구워진 치아바타는 식힘망 위로 옮겨 식혀준다.

13　Transfer the bread to the stone; slip it gently with parchment.

14　Pour the boiling water on nicely hot pebble with kettle then quickly close the oven door to trap the steam.
TIP. Be sure to wear oven gloves when you started because steam from pebbles can cause a scald.

15　When using a convection oven, because the surface of the dough dries quickly by wind; so turn off the oven for 5 minutes. Even if the oven is turned off, the dough is baked by a sufficiently heated stone plate.

16　Turn the oven on after 5 minutes, lower the oven temperature to 445°F then bake the ciabatta approximately 7 to 8 minutes. Remove the ciabattas from the oven, and cool completely on a wire rack.

VEGAN OLIVE CIABATTA

올리브 치아바타

프랑스 남부에서 생산되는 프로방스 허브와 어떻게 먹어도 맛있는 블랙 올리브를 치아바타에 듬뿍 넣는 레시피예요. 향긋하고 질 좋은 올리브유를 반죽에 넣어 빵을 만들고 있으면 따뜻한 프랑스 남부 지역에 있는 기분이 든답니다. 날씨가 좋은 주말, 리프레시하고 싶을 때 만들어보세요. 올리브 치아바타는 가벼운 식사용으로도 와인에 페어링하기에도 좋은 만능 빵이에요.

Provence Herb is commonly produced in southern France, and will be used in this recipe. Try baking ciabattas by using high-quality olive oil and black olives. Infuse the dough with Provence Herbs, and you will be able to experience the warmth of southern France in your own home. I recommend making this bread whenever you want to feel a little special on a nice weekend. Olive ciabatta goes well with meals or wine snacks. It's an all-rounder of breads.

8개의 치아바타   8 ciabattas

Ingredient

반죽 재료

트레디션 밀가루 200g
강력분 300g
물 340g
소금 10g
드라이이스트 2g
고생지 100g(생략가능)
바시나주(★반죽 후반부에 넣는 물) 40g
올리브유 50g

충전물

올리브 슬라이스 120g
이탈리안 허브 2g

Dough

French bread flour (tradition française) 200g
High gluten flour(gruau flour) 300g
Water 340g
Salt 10g
Dry yeast 2g
Pâte fermentée 100g
Bassinage water 40g
Olive oil 50g

Filling

Sliced olives 120g
Italian herb 2g

Pre-Check

주의 사항 P.035

Ⓐ + Ⓑ + Ⓒ + Ⓘ
필독!

Notice P.035

Ⓐ + Ⓑ + Ⓒ + Ⓘ
read the suggestion!

Recipe Timeline

한눈에 보는 레시피 타임라인

① 반죽하기
Mixing

③ 45분 후 사방접기
Bulk(45min), Folding

④ 다시 45분 발효(1차 발효 총 90min)
Bulk(45min)

⑤ 냉장고에 넣기 전 사방접기
Folding

⑥ 다음 날까지 냉장고에서 1차 발효
bulk fermentation in the fridge overnight

⑦ 긴 사각형으로 분할하기
Dividing and Panning

⑧ 250℃로 오븐 예열하기(돌판 넣고 예열 그동안 2차 발효 1시간)
Pre-heat the oven (with baking stone) during Proofing(60min)

⑨ 굽기(250℃, 오븐 끄고 5min)
Baking(480°F/250℃, 5min oven off)

⑩ 5분 후 다시 230℃로 오븐 켜서 8분 더 굽기
445°F/230℃, 8min oven on)

⑪ 식히기
Cooling

Pre-Cook

미리 준비할 것

1. 충전물 : 올리브는 물기를 뺀 후 준비하고, 이탈리안 허브와 섞어둔다.
2. 사각 용기 : 치아바타는 별도의 성형 없이 네모난 형태로 잘라서 굽는 빵이므로 반죽을 담을 사각 용기를 사용하면 좋다.

1. Filling : Drain the water from the sliced olives before using.
2. Rectangular dough container : There is no shaping for ciabatta, all you have to do is divide the dough into squares. Using rectangular-types helps to form a square.

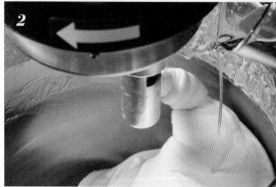

Let's Bake 1

1 믹싱 볼에 물을 먼저 넣고 바시나주, 올리브유를 제외한 나머지 반죽 재료를 넣는다. 재료가 균일하게 섞일 수 있도록 저속으로 3분 믹싱하고, 중속으로 6~7분 더 믹싱한다.

2 반죽 표면에 윤이 나고 얇은 글루텐이 생기면, 믹싱기를 중속으로 맞추고 바시나주와 올리브유 순으로 소량씩 천천히 부어준다.

TIP. 분량의 바시나주와 올리브유를 한번에 넣으면 반죽이 믹싱 볼에서 미끄러질 수 있다. 반죽에 물과 올리브유가 서서히 흡수될 수 있게 소량씩 부어주는 것이 중요하다.

3 반죽이 매끄러워지면 믹싱을 멈추고 올리브 슬라이스를 넣은 후 반죽과 올리브가 고르게 섞일 때까지 저속으로 믹싱한 후 다시 중속으로 속도를 올려 반죽이 믹싱 볼에서 떨어질 때까지 믹싱한다.

1 Pour the water into a mixing bowl then add the remaining dough ingredients except bassinage and olive oil. Knead for 3 minutes on low speed then about 6 to 7 minutes on medium speed.

2 If your dough gets smooth and glossy and the gluten strands have developed such like photo, add water and olive oil just a little bit at a time on medium speed.

TIP. If you add lots of water or oil at once, dough spins in the bowl. It's important to pour in a small amount of water and olive oil into the dough so that it can absorbed slowly.

3 When the dough gets smooth, add the sliced olives and run the mixer on low speed until evenly combined.

4   1차 발효는 45분씩 두 차례, 총 90분간 진행한다. 올리브유를 바른 사각 용기에 반죽을 붓고 사방접기 해 반죽을 조심스럽게 뒤집은 후 45분간 1차 발효를 시작한다.

5   45분 후 사방접기 한다. 반죽을 조심스럽게 뒤집어놓고 다시 45분간 발효한다.

6   1차 발효 후 반죽을 다시 사방접기 해 표면에 올리브유를 바르고 비닐을 얹는다. 12~18시간 동안 냉장고(3℃)에서 천천히 발효할 수 있도록 한다.

    TIP. 냉장고가 가득 차 있다면 냉장 효율을 고려해 제시된 온도보다 냉장고 온도를 더 낮춘다. 꽉 찬 냉장고의 실제 내부 온도는 설정 온도보다 높을 수 있기 때문이다.

7   다음 날 냉장고에서 반죽을 꺼내 실온에 둔다. 여름철에는 30분, 겨울철에는 1시간가량 둔다.

4   The first fermentation runs twice for 45 minutes, a total of 90 minutes. Transfer the dough to a rectangular dough container greased with olive oil. Then gently lift and fold dough over onto itself. Repeat it four sides. Then flip the dough over so the top is now the bottom. Let it rise for 45 minutes.

5   After 45 minutes, fold the four sides of dough and flip the dough over gently. Let it rise another 45 minutes.

6   After bulk fold the dough like just before then apply olive oil on the surface and cover the dough with plastic. Without lid, store in a 37℉ refrigerator overnight(12-18 hours) let it ferment slowly in the refrigerator.

    TIP. If the refrigerator is full, the actual internal temperature may be higher than the set point temperature. In that case, set the temperature of the refrigerator lower than the suggested temperature.

7   Remove the dough from the fridge, let the dough take the heat. Leave it for 30 minutes in summer and about 1 hour in winter.

8 반죽과 작업대 위에 밀가루를 넉넉히 뿌리고 반죽이 들어 있는 통을 뒤집어 자연스럽게 반죽이 통에서 분리될 때까지 기다린다. 반죽이 통에서 완전히 분리되면 손에 들러붙지 않도록 반죽 위에 밀가루를 뿌린다.

9 두께가 균일해지도록 손바닥으로 반죽을 가볍게 두드리며 폭 26cm, 길이 40cm 크기의 직사각형으로 펴준다.

10 26cm 쪽은 2등분, 40cm 쪽은 4등분해 13×10cm 크기의 작은 직사각형 8개로 분할하고 반죽 윗면이 위로 향하도록 천 위에 놓아준다.

TIP. 천 위로 반죽을 옮길 때 분할한 모양이 변형되지 않도록 주의한다.

8 Uncover and sprinkle lots of flour on top of the dough, then turn it out onto a liberally floured work surface. Wait for the dough to fall out of the container intactly. When the dough is completely separated from the container, sprinkle flour on top so that the dough no longer sticks to the hand.

9 Gently deflate the dough with your hands into a 10"×16" rectangle and make it even thickness.

10 Divide the 10 inch side in half and the 16 inch side into four divide into eight small rectangles, each 5" long × 4" wide. Laying them down on the couche the top of the dough faces upward.

TIP. Handling the dough gently, transfer each piece intactly on the couche.

Let's Bake 3

11   반죽 표면이 마르지 않도록 실온에서 1시간 동안 2차 발효한다.

     TIP. 실내가 건조하다면 반죽 위에 천을 덮어 마르지 않도록 한다.

12   2차 발효가 시작되면 오븐에 돌판과 자갈을 넣고 250℃로 예열하고, 발효가 끝나기 5분 전에는 주둥이가 긴 주전자에 물을 끓인다.

     TIP. 가지고 있는 오븐의 사양이 각기 다르므로 최고온도로 예열하면 된다.

13   2차 발효가 끝나면 반죽의 아랫면이 위로 향하도록 베이킹 시트에 일정한 간격으로 놓고 반죽에 밀가루를 가볍게 뿌린다.

     TIP. 베이킹용 철판이나 넓은 나무판 위에 테프론 시트나 베이킹 시트를 놓고 반죽을 올리면 오븐에 밀어 넣기 편하다.

11   Keeping the surface moist then let the dough to rise for 1 hours at room temperature.

     TIP. To prevent the top forms a skin, cover loosely with a floured couche.

12   Towards the start of the rising time, place a baking stone in the middle of the oven then place a pan filled with pebbles below the stone preheat the oven to 480°F. Boil water in a long-mouth kettle about 5 minutes before fermentation is completed.

     TIP. Since the specifications of the oven are all different, preheat to the highest temperature of each oven.

13   At the end of proof, place top of the dough facing up on a baking sheet, leaving a space between them. You may lightly flour on a dough.

     TIP. Using a Teflon or baking sheet on a baking pan or wooden plate for proof, make it easier to push the dough into the oven.

14	반죽이 놓인 베이킹 시트를 돌판 위에 밀어 넣는다.
15	주둥이가 긴 주전자에 담긴 뜨거운 물을 예열된 자갈 위에 붓고 재빨리 문을 닫는다. TIP. 자갈 위에 올라오는 수증기로 화상을 입을 수 있으므로 반드시 오븐용 장갑을 끼고 작업한다.
16	컨벡션 오븐을 사용할 경우 바람에 의해 반죽 표면이 빠르게 마를 수 있어 반죽을 넣자마자 5분간 오븐을 꺼준다. 오븐을 꺼도 충분히 달궈진 돌판이 있어 반죽이 익기 시작한다.
17	5분 후 다시 오븐을 켜고 230℃로 맞춘 후 7~8분 더 굽는다. 다 구워진 치아바타는 식힘망 위로 옮겨 식혀준다.

14	Transfer the bread to the stone; slip it gently with parchment.
15	Pour the boiling water on nicely hot pebble with kettle then quickly close the oven door to trap the steam. TIP. Be sure to wear oven gloves when you started because steam from pebbles can cause a scald.
16	When using a convection oven, because the surface of the dough dries quickly by wind; so turn off the oven for 5 minutes. Even if the oven is turned off, the dough is baked by a sufficiently heated stone plate.
17	Turn the oven on after 5 minutes, lower the oven temperature to 445°F then bake the ciabatta approximately 7 to 8 minutes. Remove the ciabattas from the oven, and cool completely on a wire rack.

WALNUT CRANBERRY CAMPAGNE BREAD

호두 크랜베리 깜빠뉴

빵을 배우기 전에는 맛있는 바게트나 깜빠뉴를 근처에서 구하기도 쉽지 않았고 먹어본 적도 많지 않았어요. 직접 빵을 만들기 시작하면서 갓 구운 하드빵이 얼마나 맛있는지 알게 되었어요. 밀의 향긋함과 담백함이 너무 좋아서 요즘은 하드빵에 더 빠져 있어요. 파스타나 스프에 찍어 먹거나 버터와 잼을 발라서 아침식사 대용으로 활용해도 좋아요. 모양은 예쁘지 않아도 괜찮아요. 깜빠뉴라는 이름의 의미가 '시골 빵'이거든요. 이름에서 느껴지듯 투박스러워야 더 멋스러운 빵입니다.

Before I learned how to bake, it was difficult to get a well-made baguette or pain de campagne, and as a result, I haven't tried them a lot. But ever since I started to bake my own bread, the aroma that comes from biting down freshly baked hard bread was too good to be true. The savory scent of wheat and the gentle but hearty taste made me hooked on baking hard bread these days. You can enjoy them with soup and pasta, or simply eat them with jam and butter for breakfast. It doesn't matter if the appearance of the end result isn't graceful. The crudeness itself is what makes pain de campagne so attractive. It literally means "country bread" in French.

Yield	290g 4개	290 * 4 batards

Ingredient

반죽 재료

프랑스 전통 밀가루 250g
강력분 200g
호밀가루 50g
물 350g
소금 10g
저당용 드라이이스트 2g
고생지 100g(생략가능)

충전물

구운 호두 60g
건크랜베리 100g
물 45g

Dough

French bread flour(tradition Française) 250g
High gluten flour(gruau flour) 200g
Rye flour 50g
Water 350g
Salt 10g
Dry yeast 2g
Pâte fermentée 100g

Filling

Toasted walnut pieces 60g
Dried cranberry 100g
Water 45g

Pre-Check

주의 사항 P.035

Ⓐ + Ⓑ + Ⓒ + Ⓔ +
Ⓕ + Ⓖ + Ⓘ 필독!

Notice P.000

Ⓐ + Ⓑ + Ⓒ + Ⓔ +
Ⓕ + Ⓖ + Ⓘ read the suggestion!

Recipe Timeline

한눈에 보는 레시피 타임라인

① 반죽하기
Mixing

② 둥글리기
Rounding

③ 1차 발효(45min)
Bulk(45min)

④ 폴딩
Folding

⑤ 다음 날까지 냉장고에서 1차 발효
Bulk fermentation in the fridge
overnight

⑥ 다음 날 실온에 1시간 정도 두어 반
죽 온도 올리기
Warm up(60min)

⑦ 분할
Dividing

⑧ 예비 성형
Pre-shaping

⑨ 휴지(30min)
Resting(30min)

⑩ 성형
Shaping and Panning

⑪ 2차 발효(90min)
Proofing(90min)

⑫ 굽기
Baking

⑬ 식히기
Cooling

Pre-Cook

미리 준비할 것

1. **충전물** : 사용하기 하루 전 구운 호두와 건크랜베리, 물을 넣고 잘 섞은 후 사용 전까지 냉장보
관한다.

1. **Filling** : Mix all ingredients and leave to macerate for few hours. Refrigerate the
mixture before use.

Let's Bake 1

1   믹싱 볼에 모든 반죽 재료를 넣는다. 재료가 균일하게 섞일 수 있도록 저속으로 3분 믹싱하고, 중속으로 6~7분 더 믹싱한다.

   TIP. 호밀가루가 들어간 반죽은 오버믹싱되면 반죽이 늘어지고 퍼질 수 있다. 글루텐이 잡히면 바로 믹싱을 멈추는 것이 중요하다.

2   반죽 표면에 윤이 나고 얇은 글루텐(P.042)이 생기면 충전물을 넣고 저속으로 섞는다.

   TIP. 믹싱기 훅 윗부분으로 반죽이 올라와 있는지 중간중간 체크하는 것이 좋다. 충전물이 반죽에 최대한 고르게 섞일 수 있도록 올라온 반죽은 손으로 밀어 내려준다.

3   완성된 반죽을 단단하게 둥글리기 한다.

1   Mix together all of the dough ingredients. Mix for 3 minutes on low speed then about 6 to 7 minutes on medium speed.

   TIP. If you over mix your rye dough, it becomes saggy and flatten easily so when the gluten is sufficiently developed, stop the mixer.

2   If your dough gets smooth and glossy and the gluten strands have developed, add the filling and run the mixer on low speed. Occasionally check the dough is on the dough ring.

   TIP. Check the dough condition frequently When the dough climbs up the hook, you must stop the mixer and scrape the dough off the hook. Make sure knead the dough as evenly as possible.

3   Fold the sides of the dough into the center and shape the ball with an inwards rotation.

---

4   둥글리기가 끝난 반죽은 밑이 좁은 발효통에 넣고 실온에서 45분간 1차 발효하고, 1차 발
    효 후 다시 둥글리기 해 12~18시간 동안 냉장고(3℃)에서 천천히 발효할 수 있도록 한다.

    TIP. 냉장고가 가득 차 있다면 냉장 효율을 고려해 제시된 온도보다 냉장고 온도를 더 낮춘다. 꽉 찬 냉
    장고의 실제 내부 온도는 설정 온도보다 높을 수 있기 때문이다.

5   다음 날 냉장고에서 반죽을 꺼내 실온에 둔다. 여름철에는 30분, 겨울철에는 1시간가
    량 둔다.

6   반죽과 작업대 위에 밀가루를 넉넉히 뿌리고 스크래퍼를 이용해 통에서 반죽을 꺼내
    손바닥으로 평평하게 두드린다.

    TIP. 분할하기 쉽도록 일정한 두께로 만든다.

4   Transfer the dough to a dough-rising bucket with a narrow bottom. Cover the bucket, and allow the dough
    to rise about 45 minutes. After bulk fold the dough simple and flip the dough over gently. Store in a 37 ℉
    refrigerator overnight(12-18 hours) let it ferment slowly in the refrigerator.

    TIP. If the refrigerator is full, the actual internal temperature may be higher than the set point temperature. In that case,
    set the temperature of the refrigerator lower than the suggested temperature.

5   Remove the dough from the fridge, let the dough take the heat. Leave it for 30 minutes in summer and about 1
    hour in winter.

6   Uncover and sprinkle lots of flour on top of the dough, then turn it out onto a liberally floured work surface
    using plastic scraper. Gently deflate the dough with your hands.

    TIP. If you make it even thickness, it is easy to divide by the same weight.

Let's Bake 2

7   반죽을 290g씩 4개로 분할한다.

8   단단하게 둥글리기 한 후 30분간 반죽을 휴지시킨다.

      TIP. 휴지 시간을 가지면서 단단한 반죽의 글루텐 망이 느슨해져 성형하기 좋은 반죽 상태가 된다.

9   휴지가 끝난 반죽에 밀가루를 뿌리고 이음매 부분이 위쪽으로 오게 작업대에 얹은 후 손바닥으로 두드려 기포를 빼주고 균일한 두께로 만든다.

7   Divide it into four of 290g each.

8   Shape into round and firm balls and rest the pre-shaped dough about 30 minutes.

      TIP. Resting relaxes the gluten and makes the final shaping easier.

9   Transfer the dough to a lightly floured surface. Gently deflate the dough with your hands and make it even thickness.

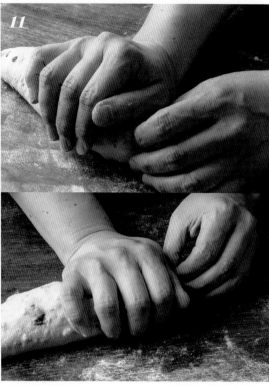

10   반죽 아래쪽을 2/3 지점까지 접고 위쪽은 1/2 지점까지 내려오도록 접어 성형한다.

11   반죽을 위에서 아래로 절반 접어 이음매를 닫는다.

10   Fold the down half up to the middle and seal and fold the top down to the middle again, slightly overlapping the first fold.

11   Fold the dough in half and seal the edges entire length.

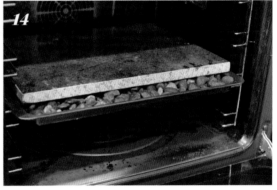

Let's Bake 3

12　이음매가 아래로 오도록 천 위에 놓는다.

　　TIP. 이때 반죽 모양을 살피고 2차 발효 시간을 조절한다. 성형한 반죽의 옆 모양은 볼록한 타원형이고 질감은 팽팽하다. 하지만 납작한 형태에 느슨한 질감이라면, 2차 발효가 제시된 시간보다 빨리 진행된 다는 것을 기억해야 한다.

13　반죽 표면이 마르지 않도록 실온에서 90분 동안 2차 발효한다.

　　TIP. 실내가 건조하다면 반죽 위에 천을 덮어 마르지 않도록 한다.

14　2차 발효 완료 1시간 전 오븐에 돌판과 자갈을 넣고 250℃로 예열하고, 발효 완료 5분 전 주둥이가 긴 주전자에 물을 끓인다.

　　TIP. 가지고 있는 오븐의 사양이 각기 다르므로 최고온도로 예열하면 된다.

12　Place them seam side down into the folds of a lightly floured couche.

　　TIP. At this time, examine the shaping and adjust the proof time. The loaves should certainly look tight and dense logs. Note that if the shaped dough is flat and loose, the final fermentation proceeds faster than the suggested time.

13　Keeping the surface moist then let the dough to rise for 90 minutes at room temperature.

　　TIP. To prevent the top forms a skin, cover loosely with a floured couche.

14　Preheat the oven to 480°F for 60 minutes before baking. Place a baking stone in the middle of the oven then place a pan filled with pebbles below the stone. Boil water in a long-mouth kettle about 5 minutes before fermentation is completed.

　　TIP. Since the specifications of the oven are all different, preheat to the highest temperature of each oven.

15   발효가 끝나면 반죽 아랫면이 위로 향하도록 베이킹 시트에 일정한 간격으로 놓는다.

16   체를 이용해 반죽 위에 밀가루를 두껍지 않고 균일하게 뿌린다.

17   면도날이나 쿠프 나이프를 이용해 빵의 세로 면을 따라 속도감 있게 쿠프를 낸다.

        TIP. 쿠프란 반죽 표면에 절개선을 넣어 가스가 쉽게 빠질 수 있도록 길을 내는 작업이다. 쿠프를 낸 반
        죽은 일정한 형태로 빠르게 팽창할 수 있다.

15   At the end of proof, using a small wooden board, turn each loaf out onto a piece of baking sheet. Top of the dough facing up and leave a space between them.

16   Using a fine-mesh sieve, dust the top of your loaves. Apply light and evenly coat.

17   Slash them with a blade(lame) or sharp knife run a single, long slash allowing the loaf to open up.

        TIP. scoring : cutting the skin of the dough to create a direction which allow the gases to escape during baking. With this step, dough can expand rapidly in expected areas and consistent manner.

18　쿠프를 넣은 반죽을 오븐 속 돌판 위에 넣는다. 예열된 자갈 위에 주둥이가 긴 주전자에 담긴 뜨거운 물을 붓고 재빨리 문을 닫는다.

　　TIP. 자갈 위에 올라오는 수증기로 화상을 입을 수 있으므로 반드시 오븐용 장갑을 끼고 작업한다.

19　반죽을 넣자마자 5분간 오븐을 꺼준다. 오븐을 꺼도 충분히 달궈진 돌판이 있어 반죽이 익기 시작한다.

20　5분 후 오븐을 켜고 230℃로 맞춘 후 18~20분간 더 굽는다. 다 구워진 깜빠뉴는 식힘망 위에서 식혀준다.

18　Carefully slip the loaves on the stone. Pour the boiling water on nicely hot pebble with kettle then quickly close the oven door to trap the steam.

　　TIP. Be sure to wear oven gloves when you started because steam from pebbles can cause a scald.

19　Turn off the oven for 5 minutes. Even if the oven is turned off, the dough is baked by a sufficiently heated stone plate.

20　Turn the oven on after 5 minutes, lower the oven temperature to 445°F then bake it approximately 18 to 20 minutes. Remove the campagne bread from the oven, and cool completely on a wire rack.

PART 2

상상 이상의 달콤함,
간식 빵

초콜릿 빵 ┃ 도넛 ┃ 브라우니

VEGAN CHOCOLATE BREAD

초콜릿 빵

논비건도 반한 진한 초코맛이 느껴지는 빵이에요. 흔히 말하는 속세의 맛 중 최고가 아닐까 싶네요. 먹어본 분마다 어떻게 비건 빵에서 이런 맛이 날 수 있냐고 놀라세요. 식감이 부드러운 빵에 쌉싸름한 초콜릿이 어우러져 입안에서 사르르 녹아요. 특히 반죽을 호두에 굴려서 굽기 때문에 고소한 맛까지 더했답니다. 아이, 어른 모두의 입맛을 사로잡을 거예요.

Non-vegans love this chocolate bread as well, and I'm pretty sure that this chocolate bread is one of the best in the world. Everyone who tried it was surprised with the taste, and even more surprised when they were told that it was vegan. The soft texture and the sweet chocolate cream melts whenever you take a bite. Not only does it have the creamy chocolate goodness in it, but also the savory, nutty aroma due to the dough being rolled on walnuts before being baked. The taste of this bread will captivate both children and adults.

Yield	시중에 판매하는 번 크기로 15개	80g * 15

Ingredient	**반죽 재료**	**Dough**
	강력분 500g	Strong flour 500g
	코코아가루 50g	Cocoa powder 50g
	소금 10g	Salt 10g
	비정제 설탕 90g	Unrefined sugar 90g
	드라이이스트 9g	Dried yeast 9g
	두유 240g	Soy milk 240g
	물 150g	Water 150g
	조청 30g	Brown rice syrup 30g
	비건 버터 80g	Vegan butter 80g
	충전물	**Filling**
	다크 초콜릿 다진 것 120g	Dark chocolate chips 120g
	토핑	**Topping**
	구운 호두 다진 것	Toasted walnut pieces
	두유 커스터드 크림	**Soy milk custard cream**
	비건 커스터드크림 60g	Vegan custard cream 60g
	두유 30g	Soy milk 30g
	코팅용 초콜릿 혼합액	**Chocolate coating**
	다진 다크 초콜릿 커버쳐 500g	Chopped dark chocolate couverture 500g
	식물성 오일 50g	Vegetable oil 50g

Pre-Check	**주의 사항 P.035**	**Notice P.035**
	Ⓐ + Ⓑ + Ⓒ + Ⓓ	Ⓐ + Ⓑ + Ⓒ + Ⓓ
	Ⓕ + Ⓗ + Ⓘ 필독!	Ⓕ + Ⓗ + Ⓘ read the suggestion!

상상 이상의 달콤함, 간식 빵

Recipe Timeline

① 반죽하기
   Mixing

② 둥글리기
   Rounding

③ 1차 발효하기(60min)
   Primary Fermentation/Bulk(60min)

④ 가스 빼기
   De-gas

⑤ 분할하기
   Dividing

⑥ 둥글리기
   Pre-shaping

⑦ 휴지하기(20min)
   Resting(20min)

⑧ 성형하기
   Shaping and Panning

⑨ 호두에 굴린 후 2차 발효하기(90min)
   Final Fermentation/Proofing(90min)

⑩ 굽기(170℃, 13min)
   Baking(340°F/170℃, 13min)

⑪ 초콜릿 코팅하기
   Coating

⑪ 냉동실에 넣어 초콜릿 굳히기
   Cooling

Pre-Cook

미리 준비할 것

1. **토핑** : 호두는 오븐 150℃에서 10분 구워 준비한다.

1. **Topping** : Prepare walnuts by roasting in a 300 °F oven for 10 minutes

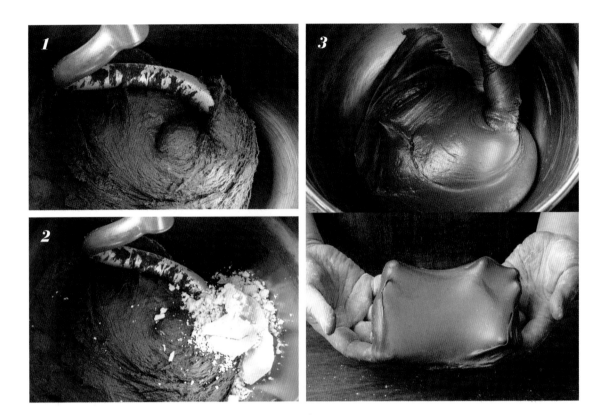

Let's Bake 1

1   믹싱 볼에 물과 두유를 먼저 넣고 비건 버터를 제외한 나머지 반죽 재료를 넣는다. 저속으로 3분 믹싱하고, 중속으로 3분 더 믹싱한다.

2   비건 버터를 넣고 중속으로 6~7분 더 믹싱한다.

3   반죽 표면에 윤이 나고, 사진과 같은 얇은 글루텐이 생기면 믹싱기를 멈춘다.

1   Pour the water and soy milk into a mixing bowl then add the remaining dough ingredients except vegan butter. Knead for 3 minutes on low speed then 3 minutes on medium speed.

2   Add vegan butter and run for 6~7 minutes on medium speed.

3   If your dough is smooth and glossy and the gluten strands have developed such like photo, stop the mixer.

상상 이상의 달콤함, 간식 빵

4   미리 준비한 초코 칩 충전물을 넣고 반죽에 고루 섞일 때까지 저속으로 믹싱한다.

5   완성된 반죽은 가장자리를 중심으로 모아 단단하게 둥글리기 한다.

6   둥글리기 한 반죽은 밑이 좁은 발효통에 넣고 실온에서 60분간 1차 발효한다.
        TIP. 반죽이 소량인 경우 둥글게 모아준 후 발효를 하면 퍼지지 않고 발효가 더 잘된다.

7   60분 발효 후 모습이다.

4   Add the chocolate chips and run the mixer on low speed until evenly combined.

5   Fold the sides of the dough into the center and shape the ball until the gluten start to tighten.

6   Transfer the dough to a dough-rising bucket with a narrow bottom. Cover the bucket, and allow the dough to rise about 1 hours.
        TIP. If your made small amount, shape the dough into a ball for Primary Fermentation the dough will rise up instead of sideways and rise correctly.

7   After 1 hours fermentation.

8   1차 발효가 끝나면 작업대와 반죽에 밀가루를 뿌리고 스크래퍼를 이용해 통에서 반죽
     을 꺼내 손바닥으로 평평하게 두드린다.

     TIP. 분할하기 쉽도록 일정한 두께로 만든다.

9   반죽을 80g씩 15개로 분할하고 둥글리기 한다.

8   At the end of the rise, lightly flour the dough, turn the dough out onto a eightly floured surface using plastic
     scraper. Gently deflate the dough with your hands.

     TIP. If you make it even thickness, it is easy to divide by the same weight

9   Divide it into 15 of 80g each and shape into round and firm balls.

10　둥글리기가 끝난 후 20분 동안 휴지시킨다. 이때 두유 커스터드 크림 재료를 고르게 섞어서 준비해둔다.

11　휴지가 끝나면 단단해지도록 다시 둥글리기 한 반죽에 두유 커스터드 크림을 발라준다.

12　구운 호두로 반죽 표면을 감싼다.
　　TIP. 발효되면서 호두가 떨어지지 않도록 손으로 꾹꾹 눌러준다.

13　2차 발효는 온도 27~28℃, 습도 70~80% 상태로 90분간 진행한다. 2차 발효 완료 10분 전에 오븐을 190℃로 예열한다.

10　Rest the pre-shaped dough about 20 minutes. During the rest prepare the soy custard cream.

11　After this rest, shape the pre-shaped dough into tight rounds. Brush all over with the soy custard cream.

12　Cover the dough with toasted walnuts.
　　TIP. Press well the surface of each dough into the walnuts. Return them to the baking sheet.

13　Allow the dough to rise for 1 1/2 hours, in a condition of 80~82°F, 70~80% to humidity. When there's 10 minutes left to proof, preheat your oven to 375°F with a rack in the center.

14 2차 발효가 끝나면 오븐 온도를 170℃로 낮추고 반죽을 오븐에 넣은 후 12~13분 굽는다. 구워진 빵은 식힘망 위에서 완전히 식힌다.

15 빵이 식는 동안 코팅용 초콜릿 혼합액을 만든다. 다크 초콜릿 커버처를 내열 용기에 담아 전자레인지에 넣은 후 가열한다. 녹인 다크 초콜릿에 식물성 오일을 넣고 섞는다.
TIP. 전자레인지로 초콜릿을 녹일 때 타지 않도록 30초 간격으로 작동시켜 녹여준다.

16 빵이 식으면 코팅용 초콜릿 혼합액에 담가 코팅한 후 냉동실에 5분간 넣어 굳힌다.

14 Towards the end of the rise time, put the pan in the oven then reduce the oven's temperature to 340°F/170°C and bake for 12 to 13 minutes. Remove from the oven and let cool completely on the pan on a rack.

15 Put the coating chocolate into a heat proof bowl or microwave safe container. Place it in microwave and melt it completely being careful not to burn. Add the vegetable oil and stirring until the mixture is evenly mixed.
TIP. To melt the chocolate by microwave, melt it gradually so it doesn't burn stop and stirring it every 30 seconds.

16 Dip the bread in melted dark chocolate mixture, then put on prepared baking sheet. Put in the freezer for 5 minutes to allow them to rest until the chocolate has set.

VEGAN DOUGHNUTS
도넛

특히 도넛 레시피는 우리가 기존에 알고 있는 특유의 부드러운 식감을 재현해내고 싶었어요. 브리오슈 느낌의 가벼운 도넛을 만들고 싶었는데 반죽을 부드럽게 만드는 동물성 재료를 사용할 수 없다 보니, 튀기고 나면 빠른 속도로 질겨지더라고요. 다양한 밀가루를 사용해보며 부드러운 식감을 찾기 위해 매일 밤 도전했답니다. 보통 달콤한 맛이 나는 빵 반죽에는 강력분을 쓰나, 원하는 식감에 따라 중력분 또는 박력분을 사용할 수 있어요. 그중 중력분으로 테스트했을 때 제 입맛에 가장 맛있고 부드러운 도넛이 나왔어요. 여러분도 원하는 식감에 맞게 밀가루를 바꿔서 만들어보실래요?

I wanted to make the texture of this donut as close as possible as the original. I wanted to make a light donut like the brioche, but since I could not use animal-based ingredients, the texture was tough after deep frying it. I tried using different kinds of flour every night to find just the right ingredients to make it soft and delicate. Trial and error repeated again and again on a daily basis, and finally I found the right balance. It was by using medium flour. Yes, bread flour is usually used for bread that tastes sweet, but you can use medium flour or soft flour depending on your preferred texture. Try experimenting with flour depending on your preference.

Yield	개당 60g, 18개	60g * 18 Doughnuts

Ingredient	**반죽 재료**	**Dough**
	중력분 500g	All-purpose flour 500g
	소금 10g	Salt 10g
	비정제 설탕 90g	Unrefined sugar 90g
	드라이이스트 10g	Dry yeast 10g
	물 130g	Water 130g
	두유 200g	Soy milk 200g
	비건 버터 130g	Vegan butter 130g
	도넛 충전 크림	**Donut filling cream**
	두유 400g	Soy milk 400g
	비정제 설탕 90g	Unrefined sugar 90g
	소금 2g	Salt 2g
	바닐라에센스 5g	Dry yeast 5g
	박력분 30g	Vanilla essence 30g
	식물성 오일 20g	Vegetable oil 20g
	다크 초콜릿 160g	Dark chocolate 160g
	튀김용 준비물	**To fry**
	식물성 기름 넉넉히	Lots of vegetable oil for frying
	온도계	Deep fry thermometer

Pre-Check	**주의 사항 P.035**	**Notice P.000**
	Ⓐ + Ⓑ + Ⓒ + Ⓓ + Ⓗ	Ⓐ + Ⓑ + Ⓒ + Ⓓ + Ⓗ
	필독!	read the suggestion!

BREADS THAT ARE SWEETER THAN YOUR IMAGINATION: DESSERT BREADS.

Recipe Timeline

한눈에 보는 레시피 타임라인

① **반죽하기**
Mixing

② **1차 발효하기(60min)**
Bulk(60min)

③ **가스 빼기**
De-gas

④ **분할(10min)**
Dividing

⑤ **둥글리기**
Rounding

⑥ **휴지(20min)**
Resting(20min)

⑦ **둥글리기(10min)**
Shaping-rounding(10min)

⑧ **2차 발효하기(90min)**
Proof(90min)

⑨ **튀기기**
Frying

⑩ **식힘망에서 식히기**
Cooling

⑪ **크림 넣기**
Fill the Doughnuts

Pre-Cook

미리 준비할 것

1. **도넛 충전 크림** : 도넛 충전 크림을 원하는 맛으로 준비하고 크림을 넣을 짤주머니를 준비한다.

2. **종이 호일** : 종이 호일을 10×10cm 크기의 정사각형으로 18개 잘라 준비한다.

1. **Doughnut Cream** : Pre-cook the donut filling cream whatever you'd like for a filling. And prepare a pastry bag for the cream.

2. **Paper sheets** : Prepare 18 of 4×4 inch square parchment paper sheets.

Let's Bake 1

1 믹싱 볼에 물과 두유를 먼저 넣은 후 비건 버터를 제외한 나머지 재료를 넣는다. 저속으로 3분, 중속으로 3분 믹싱한다.

2 비건 버터를 넣고 중속으로 6~7분 더 믹싱한다.

3 반죽 표면에 윤이 나고 사진과 같은 글루텐이 생기면 믹싱기를 멈춘다.

1 Pour the water and soy milk into a mixing bowl then add the remaining dough ingredients except vegan butter. Knead for 3 minutes on low speed then 3 minutes on medium speed.

2 Add vegan butter and run for 6~7 minutes on medium speed.

3 The gluten strands have developed such like photo, stop the mixer.

4   완성된 반죽은 가장자리를 중심으로 모아 단단하게 둥글리기 한다.

5   둥글리기 한 반죽은 밑이 좁은 발효통에 넣고 실온에서 60분간 1차 발효한다.
       TIP. 반죽이 소량인 경우 둥글게 모아준 후 발효를 하면 퍼지지 않고 발효가 더 잘된다.

6   60분 발효 후 모습이다.

7   1차 발효가 끝나면 작업대와 반죽에 밀가루를 뿌리고 스크래퍼를 이용해 반죽을 꺼
       낸다.

4   Fold the sides of the dough into the center and shape the ball until the gluten start to tighten.

5   Transfer the dough to a dough-rising bucket with a narrow bottom. Cover the bucket, and allow the dough to rise about 1 hours.
       TIP. If your made small amount, shape the dough into a ball for Primary Fermentation The dough will rise up instead of sideways and rise correctly.

6   After 1 hours fermentation.

7   At the end of the rise, turn the dough out onto a lightly floured surface using plastic scraper.

8   반죽을 각각 60g으로 분할하고 둥글리기 한 후 20분간 휴지시킨다.
     TIP. 반죽을 가볍게 눌렀을 때 탄력감이 줄고 폭신하면 휴지가 잘된 것이다.

9   휴지가 끝난 후 반죽 위에 밀가루를 뿌리고 다시 한번 둥글리기 한다. 미리 준비한 종이 호일 위에 반죽의 이음매 부분이 아래로 향하도록 놓는다.

10   2차 발효는 온도 27~28℃, 습도 70~80% 상태로 90분간 진행한다. 사진은 발효 후 반죽의 모습이다.

8   Weigh out 60g pieces each and pre-shape the pieces into balls. Rest the pre-shaped dough about 20 minutes.
     TIP. At the end of resting, you'll feel less resistance and soft when you touch the dough.

9   Rounding the dough after resting. Place the dough onto each of the parchment paper sheets; seam side down.

10   Allow the dough to rise for 1 1/2 hours, in a condition of 80~82°F, 70~80% to humidity. The picture above shows the dough after fermenting.

11 2차 발효는 온도 27~28℃, 습도 70~80% 상태로 90분간 진행한다. 사진은 발효 후 모습이다.

12 2차 발효가 끝나기 10분 전, 도넛을 튀길 준비를 한다. 우묵한 냄비에 튀김용 기름을 넣고 170℃로 데운다.
　　　TIP. 기름 온도가 필요 이상으로 높으면 반죽의 겉은 타고 속은 익지 않는다. 온도계를 꽂아두고 170℃가 넘지 않게 주의한다.

13 반죽의 한쪽 면이 밝은 갈색이 될 때까지 튀긴다(2분 정도 걸린다).
　　　TIP. 먼저 반죽 하나를 시험 삼아 튀겨 발효 상태를 체크해도 좋다. 반죽이 기름에 가라앉을 경우 발효가 덜 된 것이니 발효 시간을 조금 더 갖는다.

14 뒤집어서 다른 면도 같은 색이 날 때까지 튀긴다. 여러 번 뒤집지 않고 양면의 구움색이 같도록 튀긴다.

11 Allow the dough to rise for 1 1/2 hours, in a condition of 80~82°F, 70~80% to humidity. The picture above shows the dough after fermented.

12 When there's 10 minutes left to proof, prepare to fry the doughnuts. Heat the oil in a deep pan to 340°F.
TIP. When the oil is too hot the surface to brown too fast while the inside remains raw, Using a deep-fry thermometer heat the oil to the right temperature.

13 Carefully place the doughnuts in the oil, and fry until golden brown. Each side takes about two minutes.
TIP. You can check the stage of the fermentation by frying a dough. If the dough sink to the bottom, it's not fermented enough. Let the dough to proof a while.

14 Turn over and cook the second side; turning once to same brown color both sides.

15   튀겨진 반죽의 기름이 충분히 빠질 수 있도록 식힘망에 올려둔다.

16   젓가락을 사용해 식은 도넛 옆면에 크림이 들어갈 구멍을 낸다.

17   크림을 짤주머니에 넣어 도넛 속으로 35~40g씩 주입한다.
        TIP. 크림을 더 넣고 싶으면 제시된 크림 양보다 넉넉하게 만들어둔다.

15   Drain the oil on wired rack. Let the doughnuts thoroughly cool.

16   Poke a hole in the side of each doughnut with a chopstick to fill the cream.

17   With a pastry bag, fill the cream about 35 to 40g each into the cooled doughnuts.
        TIP. If you want extra-filled doughnuts, prepare the cream generously.

VEGAN BROWNIES

브라우니

달콤함으로 꽉 채운 꾸덕꾸덕한 브라우니는 굽자마자 살짝 따뜻할 때 바로 먹으면 좋아요. 랩으로 밀봉해 숙성시킨 후 다음 날 먹어도 좋고요. 바나나와 냉동 과일, 아가베 시럽을 넣고 블렌더로 갈아 하루 정도 얼리면 꽤 맛있는 홈메이드 아이스크림을 만들 수 있어요. 브라우니 위에 아이스크림을 듬뿍 올리고 과일이나 견과류, 메이플 시럽을 더해 즐긴다면 얼마나 좋을까요. 홈카페가 따로 없어요!

Eat them while they are warm, or eat them after sealing and aging them for a day. It doesn't matter, since brownies are always amazing. Mix bananas, frozen fruits and agave nectar into the blender and freeze it for a day or so. Place a scoop of it onto the sweet brownie and add some fruits, nuts and maple syrup on top. Try it once, and going to cafes for brownies will be a thing of the past.

Yield

17×17cm 정사각형 틀 분량

17×17cm square tin

Ingredient

믹싱 재료

다크 초콜릿 170g
코코넛 오일 80g
강력분 80g
비정제 설탕 120g
소금 1g

달걀 대체재

두유 100g
아마씨가루 10g

Mixing

Dark(non-dairy) chocolate chips 170g
Coconut oil 80g
Strong flour 80g
Unrefined sugar 120g
Salt 1g

Egg substitute

Soy milk 100g
Ground flax 10g

Recipe Timeline

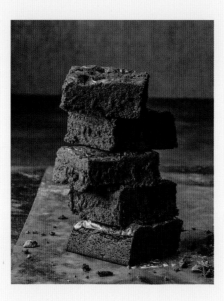

한눈에 보는 레시피 타임라인

① 오븐 예열(160℃)
Oven preheating(160℃)

② 재료 준비 및 반죽
Prep and mixing

③ 굽기(160℃, 20min)
Baking(320°F/160℃, 20min)

Pre-Cook

미리 준비할 것

1. **달걀 대체재** : 따뜻한 두유에 아마씨가루를 섞어둔다.

2. **녹인 다크 초콜릿** :

방법 ① 중탕 뜨거운 물이 든 냄비에 초콜릿이 담긴 용기를 담근 후 서서히 저어준다. 초콜릿에 물이 섞이지 않도록 하는 것이 중요하다.

방법 ② 전자레인지 30초 간격으로 작동을 멈추면서 섞어주고 다시 가열한다. 초콜릿이 타지 않도록 이 과정을 반복하며 서서히 녹여준다.

1. **Egg substitute** : Make the flax eggs by stirring the ground flax and warm soy milk in a small bowl.

2. **Melt chocolate** :

Way ① Bain Marie : Melt the chocolate in a metal bowl placed in a pot of boiling-hot water, making sure the bottom of the bowl does not actually touch the hot water. The important thing is that chocolate and water do not mix.

Way ② By microwave : Melt the chocolate in microwave-safe bowl, in 30 second intervals, stirring frequently, until the chocolate has fully melted. Remove from microwave and give it a stir frequently, melt it gradually so it doesn't burn.

Let's Bake 1

: 오븐은 반죽 전 미리 160~170℃로 예열한다.

1 달걀 대체재와 비정제 설탕, 소금을 함께 넣어 믹싱기의 비타로 저속으로 믹싱하여 완전히 녹인다.

2 녹인 다크 초콜릿을 넣어 고르게 섞는다.

3 밀가루를 체로 쳐서 믹싱 볼에 넣고 밀가루가 보이지 않을 정도로 가볍게 섞는다.

Before mixing, preheat your oven to 320~338℉.

1 Add the flax eggs and sugar, salt together, whisk until melted.

2 Add the Melted chocolate and whisk until evenly combined.

3 Over the bowl, add the sifted flour and stir until just combined.

4   정제 코코넛 오일을 넣고 섞어준다.

5   완성된 반죽을 정사각형 틀에 넣는다.

6   주걱으로 표면을 평평하게 정리한 후 예열된 오븐에 30분간 굽는다. 구워진 브라우니
는 틀 그대로 식히고, 식으면 틀에서 꺼낸다.

     TIP. 믹싱기 없이 볼과 휘퍼만 사용해서 만들 수 있다.

4   Add the coconut oil and whisk until evenly combined.

5   Pour the batter into the prepared square pan.

6   Smooth out the top with a spoon and bake for 30 minutes in preheated oven. Cool in the pan before cutting.
Once the brownies are cool, turn it out from the pan.

     TIP. You can make it with just a mixing bowl and a whipper without using a kneading machine.

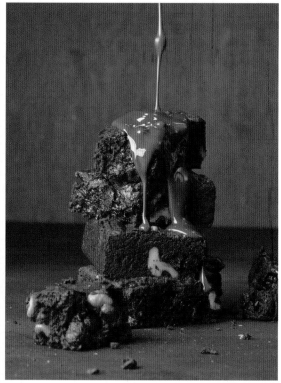

상상 이상의 달콤함, 간식 빵

PART 3

VAKE VEGAN BAKING

한 가지 반죽으로 만드는
세 가지 빵

세 가지 빵을 완성하는 만능 반죽

단팥빵 | 맘모스빵 | 인절미크림빵

HOW TO ENJOY 3 KINDS OF BREAD WITH A SINGLE DOUGH

세 가지 빵을 완성하는 만능 반죽

이 반죽 하나만 있으면 단팥빵 6개와 맘모스빵 2개, 그리고 인절미크림빵 6개를 만들 수 있어요. 다양한 추억의 빵을 만들고 싶을 때 유용한 반죽이에요. 반죽 후 1차 발효가 끝나고 분할 단계부터는 세 가지 빵에 맞게 각각 다른 레시피를 적용하니 참고해주세요. 만약 세 가지 빵 중 하나만 집중적으로 만들고 싶다면, 반죽 1배합으로 원하는 빵을 만들어도 좋아요. 그럴땐 반죽 외 충전물이나 크림의 양을 3배로 계량해 사용하면 됩니다.

With this dough, you can make six sweet red-bean breads, two mammoth breads, and six injeolmi cream breads. It is the key ingredient for various nostalgic breads. Once the first stage of fermentation is over after kneading, different recipes are used when dividing it. If you wish to focus on one of the three breads instead of making all three, that's fine as well. Just don't forget to make sure you make three times the amount of fillings or cream.

Yield	단팥빵 6개 맘모스빵 2개 인절미크림빵 6개	6 Sweet red bean paste bun(Danpatbang) 2 Mammoth bread 6 Bean cream bread(Injeolmi cream)
Ingredient	**반죽 재료** 강력분 300g 박력분 200g 두유 355g 쌀조청 20g 비정제 설탕 80g 소금 10g 드라이이스트 골드 8g 비건 버터 110g	**Dough** Strong flour 300g Cake flour 200g Soy milk 355g Brown rice syrup 20g Unrefined sugarr 80g Salt 10g Dry yeast gold label 8g Vegan butter 110g
Pre-Check	**주의 사항 P.035** Ⓐ + Ⓑ + Ⓒ + Ⓓ + Ⓗ + Ⓘ 필독!	**Notice P.035** Ⓐ + Ⓑ + Ⓒ + Ⓓ + Ⓗ + Ⓘ read the suggestion!
Recipe Timeline	**한눈에 보는 레시피 타임라인** ① 재료 준비(10min) Prep(10min) ② 반죽하기(5min) Mixing(5min) ③ 1차 발효 Proof ④ 분할/둥글리기 Dividing/Rounding ⑤ 휴지(20min) Resting(20min)	

Let's Bake 1

1   모든 반죽 재료를 믹싱 볼에 넣고 저속으로 3분, 중속으로 8~10분간 믹싱한다.
TIP. 부드러운 식감을 위해 버터를 처음부터 넣어 글루텐이 과도하게 형성되는 것을 막는다.

2   완성된 반죽의 가장자리를 중심으로 모아 둥글리기 한 후 통에 넣어 실온에서 60분간 1차 발효한다.

3   1차 발효가 끝나면 밀가루를 뿌린 작업대 위에 스크래퍼로 반죽을 꺼낸 후 손바닥으로 평평하게 두드린다.
TIP. 분할하기 쉽도록 일정한 두께로 만든다.

4   단팥빵 반죽 60g×6개, 맘모스빵 반죽 180g×2개, 인절미크림빵 반죽 60g×6개로 분할해 둥글리기 한 후 뚜껑을 덮어 20분간 휴지시킨다.

1   Mix together all of the dough ingredients. Mix for 3 minutes on low speed then about 8 to 10 minutes on medium speed.
TIP. For the soft texture add the butter from the beginning. Make sure that the gluten isn't fully developed.

2   Fold the sides of the dough into the center and shape the ball then transfer the dough to a dough-rising bucket. Allow the dough to rise about 60 minutes.

3   At the end of the rise, turn the dough out onto a lightly floured surface using plastic scraper. Gently deflate the dough with your hands.
TIP. If you make it even thickness, it is easy to divide by the same weight.

4   Divide it into 6 of 60g for sweet red bean paste bun, 2 of 180g for Mammoth bread, 6 of 60g for Bean cream bread. Pre-shape the pieces into balls. Cover them to prevent dry and rest the pre-shaped dough about 20 minutes.

VEGAN SWEET RED-BEAN BREAD 'DAN PAT BBANG'

단팥빵

한국에서 호불호 없이 모든 세대에 사랑받는 빵이라고 하면 단팥빵을 빼놓을 수 없을 거예요. 제 주변에도 단팥소를 가득 넣은 단팥빵과 우유를 같이 먹는 게 가장 좋다는 분들이 많아서 일반 단팥빵과 맛에 차이가 없는 단팥빵을 만들고 싶었어요. 한번 만들 때 많이 만들어서 소분해 냉동실에 넣어두고 생각날 때마다 하나씩 꺼내 먹으면 얼마나 든든한지 몰라요.

This is one of the most popular breads in Korea that is loved by all generations. Since many of my friends have said that their favorite bread was this, I did my best to make it taste and feel as close as possible to the original. Make dozens of these and put them in the freezer one by one.

Yield	6 buns	6 buns

Ingredient	**반죽 재료** 만능 반죽 레시피(P.151)로 60g씩 분할해 둥글리기 한 반죽 6개	**Dough** To see the dough recipe(P.151) 60G rounded dough 6 pieces
	단팥소 저당용 통팥앙금 300g 두유 15g 비건 버터 10g 구운 호두 분태 40g	**Sweet red bean paste** Low sugar red bean paste 300g Soy milk 15g Vegan butter 10g Toasted walnut pieces 40g
	두유액 두유 30g 아가베 시럽 15g	**Soy milk wash** Soy milk 30g Agave syrup 15g
	장식용 검은깨 적당량	**Topping** Black sesame handful seeds

Pre-Check	**주의 사항 P.035** Ⓕ + Ⓗ + Ⓘ 필독!	**Notice P.035** Ⓕ + Ⓗ + Ⓘ read the suggestion!

한 가지 반죽으로 만드는 세 가지 빵

Recipe
Timeline

한눈에 보는 레시피 타임라인

① 단팥소 만들기(1시간 이상 굳히기)
Making red bean paste(60min)

② 반죽 레시피(P.151)
Mixing Recipe(P.151)

③ 성형 - 팥앙금 싸기
Shaping - Filling

④ 목란으로 누르기
Press

⑤ 2차 발효
Proof

⑥ 두유액 바르기
Soymilk wash

⑦ 검은깨 찍기
Finish

⑧ 굽기
Baking

Pre-Cook

미리 준비할 것

1. **단팥소** : 비건 버터(P.026)를 전자레인지로 10초간 녹인다. ⇨ 녹은 버터와 남은 재료를 섞는다. ⇨ 냉장고에 넣어 1시간 이상 굳힌다. ⇨ 60g씩 분할해 둥글게 빚은 후 사용 전까지 냉장실에 보관한다.

1. **Sweet red bean paste** : Melt the vegan butter by the microwave for about 10 seconds. ⇨ Combine melted vegan butter with the rest of ingredients. ⇨ Refrigerate the mixture for more than an hour. ⇨ Divide for 60g and shape into round balls and refrigerate before use it.

Let's Bake 1

1 휴지가 끝난 반죽과 작업대에 가볍게 밀가루를 뿌린 후 이음매 부분이 위로 향하게 작업대에 얹는다.

2 밀대로 밀어 지름 10cm의 원을 만든다.

3 준비해놓은 단팥소를 밀어놓은 반죽 위에 하나씩 올린다.

1 At the end of the rest, turn the dough out onto a lightly floured.

2 Roll one piece of dough into a 4" circle.

3 Place prepared sweet red bean paste on the center of the circle.

4   반죽 가장자리를 중심으로 모아 앙금소를 감싼다.

5   감싼 부분을 아래 방향으로 팬닝한다.

4   Bring the sides of the dough to cover the filling.

5   Keeping the seam side down.

Let's Bake 2

6   목란을 힘껏 눌러 반죽 중앙에 넓은 구멍을 만든다.

7   2차 발효는 온도 28℃, 습도 70~80% 상태에서 90분간 진행한다. 사진은 2차 발효가 끝 난 직후의 모습이다.

8   2차 발효가 끝나기 10분 전 오븐을 200℃로 예열한다. 2차 발효 후 반죽에 두유액을 발 라주고 검은깨를 얹는다.

　　　TIP. 물기 있는 검지손가락에 검은깨를 묻혀 반죽에 찍어준다.

9   오븐 온도를 180℃로 낮추고 12~13분 굽는다.

6   press the center of the dough; make a small hole with rounded wood stick or special wooden tool for this bread.

7   Allow the dough to rise for 90 minutes, in a condition of 80~82°F, 70 to 80% to humidity. The picture shows the shape after the final fermentation.

8   When there's 10 minutes left to proof, preheat your oven to 390°F. At the end of proof brush each bun with soy milk mixture and decorate with the black sesame seeds.

　　　TIP. Put water on your index finger, dip black sesame seeds, and put it on the bun.

9   Reduce the oven temperature to 355°F and bake the bread for 12 to 13 minutes.

VEGAN MAMMOTH BREAD

맘모스빵

여러 맛이 조화를 이루는 맘모스빵은 레트로 붐으로 최근 다시 유행하고 있는 추억의 빵이에요. 맘모스빵만 찾아다니는 빵지순례도 있다고 하니 말이죠. 제 레시피는 팥앙금과 완두앙금을 감싼 반죽에 포슬포슬한 비건 소보로를 얹어 구워내고 달콤한 코코넛 크림과 상큼한 라즈베리 잼을 샌딩하는 거예요. 추억을 불러오면서 촌스럽지 않게 재해석했어요. 코코넛 크림과 라즈베리 잼은 다른 빵에도 잘 어울리니 사용하고 남은 것은 보관해두었다가 빵에 발라 먹거나 비건 스콘에 곁들여보세요.

The name itself represents this bread very well. A rustic look with lots of fillings, and a huge nostalgia for many people in Korea. This bread is so popular that there are people going around the country looking for the best bakeries in Korea. My recipe is to bake the dough wrapped in sweet red bean fillings and sweet pea fillings, and soft vegan soboro (peanut-flavored streusel) on top. After baking it, sandwich it with sweet coconut cream and sour raspberry jam. Leftover coconut cream and raspberry jam goes well with other breads, so don't make it go to waste.

2개

2 Sandwichs

Ingredient

반죽 재료

만능 반죽 레시피(P.151)로
180g씩 분할해 둥글리기 한 반죽 2개

Dough

To see the dough recipe(P.151)
180G rounded dough 2 pieces

충전물

라즈베리 잼 (레시피 P.030 참고)
팥앙금 180g
(시중 제품 중 저당 앙금 사용)
완두앙금 180g

Filling

Raspberry jam (P.030)
Low sugar red bean paste 180g
Sweet pea paste 180g

코코넛 크림

냉장한 코코넛 크림(고형분) 300g
비정제 설탕 30g

Coconut cream

Hardened coconut cream 300g
Unrefined sugar 30g

두유액

두유 30g
아가베 시럽 15g

Soy milk wash

Soy milk 30g
Agave syrup 15g

소보로

Crumble

Pre-Check

주의 사항 P.035

ⓗ + ⓘ 필독!

Notice P.035

ⓗ + ⓘ read the suggestion!

한눈에 보는 레시피 타임라인

① **하루 전 라즈베리 잼 만들기(P.030)**
Making the jam previously(P.030)

② **코코넛 크림 휘핑해 냉장 보관**
Whip the cream and refrigerate it

③ **비건 소보로, 팥앙금, 완두앙금 준비
하기**
Making the crumble and preparing
the bean pastes

④ **반죽 레시피(P.151)**
Mixing(P.151)

⑤ **휴지**
Resting

⑥ **성형 후 타원형으로 밀기**
Shaping - pressing

⑦ **두유액 바르고 소보로 올리기**
Soymilk wash and put the crumble on top

⑧ **2차 발효**
Proof

⑨ **굽기**
Baking

⑩ **식힌 후 잼과 코코넛 크림 바르고 샌딩**
Assemblage

미리 준비할 것

1. **라즈베리 잼** : 최소 하루 전 잼(P.030)을 만든다.
2. **코코넛 크림** : 하루 전 코코넛 크림을 냉장고에서 충분히 굳힌 후 고형분을 뜬다. ⇨ 비정제
설탕을 넣어 휘핑한 후 크림을 만들어 사용 전까지 냉장 보관한다.
3. **비건 소보로** : 레시피(P.033)를 참고해 비건 소보로 가루를 만든다.
4. **팥앙금, 완두앙금** : 180g씩 분할해 둥글게 빚은 후 사용 전까지 냉장 보관한다.

1. **Raspberry jam** : Make the jam(P.030) in advance.
2. **Coconut cream** : Chilling can overnight in the refrigerator to harden and scoop the
coconut cream off the top of the can into a mixing bowl and add sugar. ⇨ Fluff up
the coconut cream and store in the refrigerator until use.
3. **Vegan crumble** : Make the crumble(P.033) in advance.
4. **Bean pastes** : Divide for 180g and shape into round balls and refrigerate before use.

Let's Bake 1

1   휴지가 끝난 반죽과 작업대에 가볍게 밀가루를 뿌린 후 이음매 부분이 위로 향하게 작업대에 얹는다.

2   밀대로 밀어 지름 15cm의 원을 만든다.

3   준비해놓은 팥앙금과 완두앙금을 밀어놓은 반죽 위에 하나씩 올린다.

4   반죽 가장자리를 중심으로 모아서 소를 감싼다.

1   At the end of the rest, turn the dough out onto a lightly floured. work surface.

2   Roll one piece of dough into a 6" circle.

3   Place prepared red bean paste and sweet pea paste on the center of the circle.

4   Bring the sides of the dough to cover the filling.

Let's Bake 2

5 밀대로 밀어 좁은 쪽 지름은 15cm, 넓은 쪽 지름은 23cm 정도의 타원형을 만든다.

6 감싼 부분을 아래 방향으로 팬닝한 후 반죽에 두유액을 골고루 발라주고 만들어둔 소보로를 반죽 위에 올린다.

5 Place the dough onto work space with seam side down, Roll out the dough into oval, make the shortest part to 6" and longest part to 9".

6 Place the dough on the pan with seam side down, brush each bun with soy milk mixture evenly then top with the crumble.

7   2차 발효를 온도 28℃, 습도 70~80% 상태로 90분간 진행한다. 2차 발효 완료 10분 전 컨벡션 오븐을 200℃로 예열한다. 발효가 끝나면 반죽 위의 소보로가 떨어지지 않도록 스프레이로 물을 뿌린다.

8   오븐 온도를 180℃로 낮추어 15분 동안 굽고 다 구워지면 식힘망에 옮겨서 완전히 식힌다.

9   구워진 빵의 매끈한 면에 코코넛 크림을 바르고 다른 한쪽에는 라즈베리 잼을 바른다. 다 바른 후 서로 맞닿도록 포갠다.

7   Allow the dough to rise for 90 minutes, in a condition of 80~82°F, 70 to 80% to humidity. When there's 10 minutes left to proof, preheat your oven to 390°F. At the end of proof spraying water over the dough to stick the crumble well.

8   Reduce the oven temperature to 355°F and bake the bread for 15 minutes. Remove the bread from the oven, and cool completely on a rack.

9   Spread the whipped coconut cream on one bottom of the bread and spread the raspberry jam on the other. And make two sides are sandwiched together.

VEGAN INJEOLMI CREAM BREAD
(BEAN CREAM BREAD)

인절미크림빵

크림빵을 사랑하고 인절미를 좋아하는 분이라면 반할 수밖에 없는 빵이에요. 달콤하고 풍성한 느낌의 두유 크림에 고소한 콩가루를 더해 한입 베어 물면 입안이 고소함으로 꽉 채워져요. 인절미 콩가루를 입가에 잔뜩 묻히며 먹다 보면 웃음이 절로 날 거예요.

If you love cream bread and also enjoy injeolmi (injeolmi: Korean rice cake with soybean powder on top), I promise you will fall in love with this one. Once the sweet fresh soybean cream and savory soybean powder touches your taste buds, you won't be able to forget the taste. Oh, but you will probably forget the powder on your smiling lips. Be sure to lick them off after every bite.

Yield

6개	6 breads

Ingredient

반죽 재료
만능 반죽 레시피(P.151)로
60g씩 분할해 둥글리기 한 반죽 6개

Dough
To see the dough recipe(P.151)
60g rounded dough 6 pieces

충전물
비건 인절미 크림(레시피 P.029)

Filling
Vegan injeolmi cream(P.029)

토핑
팥빙수용 콩가루 넉넉히

Topping
Sweetened soy bean powder for
patbingsu(snowflake shaved ice)

두유액
두유 30g
아가베 시럽 15g

Soy milk wash
Soy milk 30g
Agave syrup 15g

크림액
인절미 크림 25g
두유 15g

Cream wash
Vegan injeolmi cream 25g
Soy milk 15g

Pre-Check

주의 사항 P.035
Ⓗ + Ⓘ 필독!

Notice P.035
Ⓗ + Ⓘ read the suggestion!

Recipe
Timeline

한눈에 보는 레시피 타임라인

① **인절미 크림 만들기**
Making injeolmi cream

② **반죽(P.151)**
Mixing (P.151)

③ **1차 발효**
bulk

④ **분할 - 60g**
Dividing - 60g

⑤ **둥글리기**
Pre-shaping

⑥ **휴지하기**
Resting

⑦ **성형하기**
Shaping

⑧ **2차 발효**
Proof

⑨ **굽기**
Baking

⑩ **식히기**
Cooing

⑪ **인절미 크림 주입**
Piping the cream

⑫ **크림액 바르기**
Cream wash

⑬ **콩가루 묻히기**
Coat with soybean powder

Pre-Cook

미리 준비할 것

1. **비건 인절미 크림 :** 하루 전날 레시피(P.029)를 참고해 비건 인절미 크림을 만든다.

1. **Vegan injeolmi cream :** Make the cream(P.029) in advance.

Let's Bake 1

1   휴지가 끝난 반죽은 다시 단단하게 둥글리기 한다.

2   이음매 부분이 아래로 향하도록 팬닝하고 온도 28℃, 습도 70~80% 상태에서 90분간 2차 발효한다.

3   2차 발효 완료 10분 전 오븐을 200℃로 예열한다. 발효가 끝난 반죽에 두유액을 골고루 바른다.

1   Shape the dough into a tight ball after resting.

2   Place the dough seam side down onto a baking pan. Allow the dough to rise for 90 minutes, in a condition of 80~82°F, 70 to 80% to humidity.

3   When there's 10 minutes left to proof, preheat your oven to 390°F. Brush the buns with the soy milk wash substitute evenly.

4 오븐에 넣고 온도를 180℃로 낮춘 후 12~13분 굽는다.

5 구워진 빵이 완전히 식은 후 크림이 들어갈 수 있도록 옆부분에 구멍을 낸다.

6 미리 준비한 인절미 크림을 짤주머니에 담아 원하는 만큼 빵 안으로 주입한다.

4 Put the pan in the oven. Reduce the oven temperature to 355°F and bake the bread for 12 to 13 minutes.

5 Remove the buns from the oven, and cool completely on a rack. Poke a hole in the side of each bun and wiggle with a chopstick to fill the cream generously.

6 Pipe the soybean cream into the bun as you desire.

Let's Bake 2

7   크림을 충전한 빵 겉면에 크림액을 바른다.

8   토핑용 콩가루를 듬뿍 묻힌다.
      TIP. 팥빙수용 콩가루를 사용하면 단맛이 증폭된다.

7   Apply the cream wash over the entire surface of cream filled bun.

8   Coat it liberally with the sweetened soybean powder.
      TIP. Roasted soybean powder used in snowflake shaved ice is normally sweetened. So using it can add extra sweetness to the bread.

PART 4

VAKE VEGAN BAKING

비건 크루아상 반죽으로
만드는 여섯 가지 빵

비건 크루아상 반죽

크루아상 | 크러핀 | 빵 오 쇼콜라

빵 오 크랜베리 | 올리브 타프나드 & 튀긴 양파 페스츄리

과일 & 비건 크림 페스츄리

VEGAN CROISSANT DOUGH

비건 크루아상 반죽

크루아상도 비건으로 만들 수 있다고? 버터 없이는 만들 수 없다고 생각하던 크루아상도 비건 버터만 있으면 논비건 크루아상처럼 만들 수 있어요. 특히 이 반죽 하나만 있으면 크루아상 뿐만 아니라 크러핀, 페스츄리, 빵 오 쇼콜라까지 도전해 볼 수 있답니다.

Who said that croissants can't be vegan? With vegan butter, you can make croissants that taste the same as the original. Just by using this dough, you can make not only croissants, but also cruffins, pastries, and even pain au chocolats.

반죽 재료

강력분 500g

소금 10g

비정제 설탕 70g

비건 버터 70g

두유 100g

물 150g

조청 10g

세미 드라이이스트 골드 13g

고생지 100g(생략가능)

페스츄리 뚜라주용(접기용) 버터

레시피 참고(P.027)

15×15cm(140g) 크기 2개

Dough

Strong flour 500g

Salt 10g

Unrefined sugar 70g

Vegan butter 70g

Soy milk 100g

Water 150g

Brown rice syrup 10g

Semi-dry yeast gold label 13g

Fermented dough (pâte fermentée) 100g

Lamination butter

Dough recipe(P.027)

Two butter blocks in 6" square size

주의 사항 P.035

Ⓐ + Ⓑ + Ⓒ + Ⓓ 필독!

Notice P.035

Ⓐ + Ⓑ + Ⓒ + Ⓓ read the suggestion!

한눈에 보는 레시피 타임라인

① **반죽**
Mixing

② **둥글리기**
Rounding

③ **1차 발표(총 40min)**
Primary Fermentation/Bulk(40min)

④ **반죽 준비**
Preparing the dough

⑤ **뚜라주(접기)**
Roll out and fold the dough(tourage)

Let's Bake 1

1 모든 반죽 재료를 넣고 저속으로 10분, 중속으로 6~8분간 믹싱한다.

TIP. 크루아상 반죽은 수분이 적은 반죽이므로 저속으로 오래 믹싱해서 반죽을 부드럽게 한 후 중속으로 믹싱한다. 믹싱 시간이 길면 마찰열로 반죽 온도가 올라간다. 믹싱 볼과 훅을 냉동 보관해서 사용한다면 반죽 온도를 맞추는 데 도움이 된다. 또 글루텐이 부드러워지기 전까지 반죽이 훅 밖으로 겉돌 수 있으므로 중간에 확인해 반죽을 훅 쪽으로 밀어 넣어주면 좋다.

2 완성된 반죽은 500g씩 두 덩어리로 분할해 단단하게 둥글리기 한다.

1 Place entire dough ingredients in a mixing bowl and knead for 10 minutes on low speed then 6 to 8 minutes on medium speed.

TIP. The croissant dough is low hydrated and firm dough. So extend the low speed mixing period to make the dough soft enough and change the mixing speed to medium. Since the intense and long mixing time, it generates friction heat a lot. If your refrigerator is large enough, chilling the mixing bowl and hook can help to maintain the proper dough temperature. Stiff dough rotates out of the hook until the gluten softens. When the ingredients combined and formed a dough, push the dough toward the hook to allow proper mixing.

2 Divide the dough into 500g each and shape into round and firm balls.

Let's Bake 2

3   둥글리기 한 반죽은 발효통에 넣고 뚜껑을 덮어 실온에서 20분간 1차 발효한다.

4   사진은 20분 후 모습이다.

5   반죽에 밀가루를 가볍게 뿌리고 윗면이 아래로 가도록 작업대에 얹는다. 양 손바닥으로 가볍게 눌러 세로로 긴 타원형으로 만든다.

3   Place seam side down in a dough container and cover it. Allow to rise about 20 minutes at room temperature.

4   After 20 minutes fermentation.

5   Sprinkle flour on top and turn it out and deflate the dough with your hands into oval.

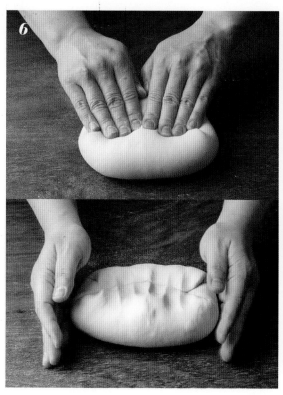

6	반죽이 너무 길어지지 않도록 주의하면서 두 번 접고 이음매를 닫아준다. 바타르(바게 트보다 두껍고 짧은 모양)로 성형해 20분 더 발효한다.

7	1차 발효가 끝난 반죽은 가볍게 밀가루를 뿌리고 반죽 윗면이 아래로 가도록 세로로 길게 작업대에 얹는다.

6	Fold the dough twice and seal the edge, shape into a short batard. Let it rise another 20 minutes.

7	At the end of bulk, transfer the dough to a lightly floured surface seam side up.

8 밀대로 밀어 40cm로 늘인다.

9 윗부분을 약간 넓게 잡고 살짝 당겨 둥근 반죽을 사각형이 되도록 2/3 지점까지 접고 아래쪽도 동일하게 접는다.

10 반죽을 90도로 회전시켜 15×30cm의 긴 사각형이 되도록 밀대로 민다.

11 반죽을 철판에 얹고 비닐로 감싼 후 딱딱하게 얼 때까지 2시간 이상 냉동시킨다.

8 Using a dough roller extend the length to 40cm.

9 Pulling the top of the dough to the side to make it angled and fold the dough into thirds.

10 Rotate the dough 90 degrees and extend the dough into a 15 x 30cm rectangle.

11 Place the dough on a pan and cover with plastic, keep it in the fridge until completely hardened, minimum 2 hours.

반죽이 단단해지면 -1℃ 온도에 12~16시간 동안 둔다. 혹은 가정용 냉장고에서 2~3 시간 동안 둔다.

TIP. 반죽이 부풀지 않고 얼지도 않게 하는 -1℃ 상태를 지키는 게 중요하다.

13 반죽을 가로로 길게 테이블에 얹고, 페스츄리 뚜라주용(접기용) 버터를 반죽 중앙에 놓는다.

14 버터가 놓여 있지 않은 양쪽의 반죽을 잘라서 버터 위에 얹은 후 빈 부분이 없도록 당겨 반죽이 서로 겹치지 않고 만나게 한다.

TIP. 반죽 → 버터 → 반죽의 층을 이루게 한다.

12 When the dough is rigid remove it from the freezer and store it in -1℃ refrigerator about 12 to 16 hours(overnight). Or refrigerate it for two to three hours at home.

TIP. At the ideal temperature, it's neither swollen nor frozen.

13 Remove the dough from the refrigerator transfer it on the work surface long horizontally. Unwrap the chilled butter block and place it in the center of the dough.

14 Cut the dough, the sides of butter. Bring the doughs on the butter and enclose the butter. Do not overlap the doughs, keep the edges straight and line them up.

TIP. It forms single layer of butter.

Let's Bake 4

15 반죽이 맞닿은 부분을 세로 방향으로 놓고 밀대를 이용해 일정한 힘으로 누른다. 버터를 반죽에 고정시키며 길이를 늘인다.

16 반죽이 40~45cm 길이가 되면 3절접기 한다. 반죽을 비닐로 감싸 20분간 냉동한 후 냉장실로 옮겨 10분간 냉각 휴지시킨다.

TIP. 냉각 휴지하는 이유는 반죽을 밀기 쉬운 상태로 만들고, 버터가 반죽에 흡수되는 것을 막아 선명한 층이 생기게 하기 위해서다.

15 Make the center seam is oriented vertically, roll out the dough lengthwise along the seam with a dough roller, pressing it downward, do.

16 When you've reached 40 to 45cm, fold the dough in thirds. Cover it with plastic. Freeze it for 20 minutes and transfer to the refrigerator and store it for 10 minutes.

TIP. Resting and chilling period makes the lamination easy and chilled butter hardly incorporates in the dough, so it can remain as a separate layers.

17 접힌 반죽의 양쪽을 칼로 자르고 자른 부분이 책 모양처럼 좌우로 위치하도록 놓은 후 세로로 길어지도록 민다.

18 반죽을 55cm로 밀어 4절접기 한다. 반죽을 다시 비닐로 감싸 20분간 냉동한 후 냉장실로 옮겨 10분간 냉각 휴지시킨다.

17 Place the dough so it looks like a book, cut the edges of the folded dough and roll out the dough lengthwise.

18 Once it's rolled out to about 55cm, do double fold. Freeze it for 20 minutes and transfer to the refrigerator and store it for 10 minutes.

VEGAN CROISSANT

크루아상

빵에 대한 열정이 가득한 날, 여유가 있는 주말, 오랜만에 생긴 휴가일에 만들어볼 만한 난도 높은 빵이에요. 크루아상을 제대로 만들기 위해서는 이틀이 걸리기 때문이죠. 제가 소개하는 대로 잘 따라 만든다면 여태껏 본 적 없는 환상적인 비건 크루아상을 맛보실 수 있을 거예요. 질감은 공기처럼 가볍고 식감은 바삭하면서 부드러워 많은 이들이 사랑하는 크루아상을 비건 버전으로 개발했습니다. 노력이 듬뿍 담긴 레시피를 소개합니다.

Croissant is loved by many people for its smooth taste and fluffy texture, and I made a vegan version of it. If you have two free days and the passion for making bread, go ahead and try making this croissant. This is a bread that requires lots of time and skill, but it will reward you once it is finished. Just keep in mind that you have to follow my recipe exactly.

Yield	고생지 사용 시 크루아상 8개 (고생지를 사용하지 않는다면 크루아상 6~7개)	8 triangles with fermented dough and 6 to 7 triangles without fermented dough for dough at your recipe.
Ingredient	**반죽 재료** 비건 크루아상 반죽(P.174) 1개 **두유액** 두유 30g 아가베 시럽 15g	**Dough** 1 Laminated croissant dough(P.174) **Soy milk wash** Soy milk 30g Agave syrup 15g
Pre-Check	**주의 사항 P.035** Ⓗ + Ⓘ 필독!	**Notice P.035** Ⓗ + Ⓘ read the suggestion!

비건 크루아상 반죽으로 만드는 여섯 가지 빵

Recipe Timeline

한눈에 보는 레시피 타임라인

① **반죽 펴기**
Final roll out

② **재단**
Cutting

③ **성형**
Shaping

④ **2차 발효(3hr)**
Final fermentation(3 hours)

⑤ **굽기**
Baking

⑥ **식히기**
Cooling

Let's Bake 1

1   접힌 반죽의 양쪽을 칼로 자르고 자른 부분이 좌우로 위치하게 놓은 후 폭 27cm, 높이 40cm 크기로 밀어준다.

        TIP. 반죽의 테두리를 잘라주면 구웠을 때 결이 더 선명해진다.

2   높이 27cm로 민 반죽을 밑변 9cm의 이등변삼각형으로 자른다.

        TIP. 재단 시 반죽이 말랑하다면 비닐을 덮고 30분간 냉장 보관해 완전히 차갑고 단단해질 때 성형하는 것이 좋다.

3   재단한 반죽의 밑변은 위로, 꼭짓점은 아래로 향하게 놓고 밑변 중앙에 1cm 크기의 칼집을 넣는다.

4   삼각형 양 끝을 옆으로 당겨서 길이를 조금 늘이고 손끝에 힘을 주어 반죽을 굴린다.

1   Making cuts at the edges of the folded dough and roll it out to 27×40cm.

        TIP. If you trim the edges of the sheeted dough, the layer is clearer when it's baked.

2   Cut into 9×27cm Isosceles triangle.

        TIP. If your dough is too soft to shape, wrap and refrigerate for 30 minutes until it gets cold and hard.

3   Arrange them so the base of triangle is facing toward you and cut a 1cm notch in the center of the base of each triangle.

4   Then tug gently outward to extend the points and widen the base a bit. Roll the base of the triangle toward the tip.

5   탄탄하게 한 바퀴 감아 코어를 단단하게 만든 후 나머지 부분은 가볍게 굴려 성형을 마무리한다.

6   삼각형의 꼭짓점이 바닥에 놓이게 해 베이킹 팬에 얹는다.
     TIP. 발효가 되면서 크기가 커지므로 반죽 간격을 여유 있게 둔다.

7   반죽이 마르지 않도록 두유·액을 발라주고 온도 27~28℃, 습도 75~80℃ 상태에서 3시간 동안 2차 발효시킨다. 사진은 2차 발효 후 모습이다.

5   Apply light pressure with your finger tips to make a tight core. Then roll them up from the bottom, trying not to roll too tightly or stretch the dough around itself.

6   Make sure the tip ends up under the bottom of the croissant, place it on a baking pan.
     TIP. They expand noticeably so leave a space between the shaped pastry.

7   Brush each croissant with the soy milk wash substitute to prevent drying out during the proof, allow the dough to rise for 3 hours, in a condition of 80~82°F, 75~80% to humidity. After final fermemtation.

Let's Bake 2

8   발효 완료 10분 전 오븐을 190℃로 예열한다. 굽기 전 두유액을 한 번 더 조심히 바른 후 오븐에 반죽을 넣고 170℃로 온도를 낮춘다.

9   13~15분간 굽고 다 구워진 빵은 완전히 식힌다.

8   When there's 10 minutes left to proof, preheat your oven to 375˚F. Delicately brush the soy milk wash substitute on top of the croissants with a soft pastry brush, put the dough in the oven. Then reduce the oven's temperature to 340˚F.

9   Bake for 13 to 15 minutes. Remove from the oven and let cool on the pan.

VEGAN CRUFFIN

크러핀

이름이 다소 생소할 수 있는 크러핀은 크루아상 반죽을 활용한 빵으로 요즘 SNS에서 많이 볼 수 있어요. 비정제 설탕에 버무린 플레인 크러핀도 맛있지만, 저는 달콤한 충전물을 채워 만들었어요. 코코넛 크림을 섞은 비건 커스터드 크림과 라즈베리 잼을 크러핀 안에 풍성하게 충전했답니다. 라즈베리 잼 대신 여러분이 좋아하는 잼을 활용한다면 더욱 즐거운 베이킹 시간이 될 거예요.

Social media has been all over cruffins for a while. As the name suggests, it is a cross between a croissant and a muffin. Plain cruffins mixed with sugar are also delicious, but I prefer having sweet fillings inside them. This one has vegan custard cream mixed with coconut cream and raspberry jam. If you have your own favorite jam, use it instead. Baking is all about enjoying the end result and the whole process itself.

Yield	8개	8 units

Ingredient

반죽 재료	**Dough**
비건 크루아상 반죽(P.174) 1개	1 Laminated croissant dough(P.174)
충전용 비건 버터	Vegan butter

두유액	**Soy milk wash**
두유 30g	Soy milk 30g
아가베 시럽 15g	Agave syrup 15g

충전물	**Filling**
라즈베리 잼(레시피 P.030)	Raspberry jam(P.030)
비건 커스터드 크림(레시피 P.030)	Vegan custard cream(P.030)

Pre-Check

주의 사항 P.035	**Notice P.035**
Ⓗ + Ⓘ 필독!	Ⓗ + Ⓘ read the suggestion!

Recipe
Timeline

한눈에 보는 레시피 타임라인

① **반죽 펴기**
Final roll out

② **재단**
Cutting

③ **성형하기**
Shaping

④ **2차 발효(2hr 30min)**
Final fermentation(2 hours 30 minutes)

⑤ **굽기**
Baking

⑥ **식히기**
Cooling

⑦ **데코레이션**
Decoration

Let's Bake 1

1   접힌 반죽의 양쪽을 칼로 자르고 자른 부분이 좌우로 위치하게 놓은 후 폭 28cm, 높이 40cm 크기로 밀어준다.

 TIP. 반죽의 테두리를 잘라주면 구웠을 때 결이 더 선명해진다.

2   28cm 길이의 폭은 2등분, 40cm 길이의 높이는 4등분해 14×10cm 크기의 사각형으로 재단한다.

3   10cm 면을 3등분해 총 24개의 사각형으로 재단한다.

1   Making cuts at the edges of the folded dough and roll it out to 28×40cm.

 TIP. If you trim the edges of the sheeted dough, the layer is clearer when it's baked.

2   Cut 28cm into 2 pieces and 40cm into 4 pieces so make four 14×10cm rectangles.

3   Then devide 10cm into 3 pieces again, so make 24 short strips in total.

Let's Bake 2

4   크러핀 틀 1개당 반죽 3개가 필요하다. 반죽끼리 3cm씩 겹쳐지도록 포갠 후 중심이 뾰족해지지 않도록 힘을 빼고 말아준다.

5   반죽을 원형 틀에 놓고 중간 부분을 눌러준다.

6   반죽이 마르지 않도록 두유액을 발라주고 온도 27~28℃, 습도 75~80℃ 상태에서 2시간 30분 동안 2차 발효시킨다. 사진은 2차 발효 후 모습이다.

4   For each cruffin, take 3 strips and stagger them for 3cm. Roll into a coil, not pointy in the middle.

5   Place it in a round mould. Put some water on your finger tip and press down the center.

6   Brush each coils with the soy milk wash substitute to prevent drying out during the proof, allow the dough to rise for 2 hours 30 minutes, in a condition of 80~82°F, 75~80% to humidity. The picture above is the appearance of the dough after the final formentation.

7 발효 완료 10분 전 오븐을 200℃로 예열한다. 오븐에 넣기 직전 손가락에 물을 묻혀
  중심 부분을 꾹 눌러준다.
  TIP. 반죽이 틀에 들어 있기에 팬에 바로 얹어 굽는 반죽보다 예열 온도가 높다.

8 반죽을 넣고 170℃로 온도를 낮춘 후 23분간 굽는다.

9 구워진 빵은 틀에서 분리해 완전히 식힌다.

10 빵 바닥에 구멍을 내고 비건 커스터드 크림과 라즈베리 잼을 짜 넣는다.
   TIP. 충전물을 넣는 대신 빵 겉면에 녹인 코코넛 오일을 발라 살짝 굳힌 후 비정제 설탕에 굴려도 좋다.

7 When there's 10 minutes left to proof, preheat your oven to 390°F. When cruffins are ready to be baked, put
  some water on your finger tip and press down the center again.
  TIP. Raise the oven temperature to preheating, if you bake the dough in the mould.

8 Place the pan into the oven and reduce the heat to 375°F and bake for 23 mins.

9 Remove from the oven and turn the cruffins out of the pan onto a rack to cool completely.

10 Pipe the cream and jam into the cruffin.
   TIP. Instead of fill the cream, you may coat the cruffin in sugar. Brush the melted coconut oil on it and let them set a while
   then coating in sugar.

VEGAN PAIN AU CHOCOLAT

빵 오 쇼콜라

크루아상 반죽에 다양한 재료를 넣어 구워볼 수 있지만 가장 인기 많은 재료는 단연 초콜릿일 거예요. 프랑스 빵집의 비에누아즈리 코너에서 꼭 찾아볼 수 있는 빵 중 하나로 네모나게 자른 크루아상 반죽에 달콤한 초콜릿 바를 넣고 둥글게 말아주면 누구에게나 사랑받는 아침 식사 메뉴가 탄생합니다.

There are so many ingredients that you can put inside a croissant, but the most popular one is chocolate. Go to a French bakery, and you will most likely find one of these in the Viennoise section. To make one, cut the croissant dough into square shapes and roll them after placing chocolate bars in them. They make for a great snack or a quick breakfast meal.

Yield

8개

8 rolls

Ingredient

반죽 재료

비건 크루아상 반죽(P.174) 1개

충전용 비건 버터

충전물

초코 스틱 16개

당절임한 오렌지 스틱 8개

두유액

두유 30g

아가베 시럽 15g

Dough

1 Laminated croissant dough(P.174)

Vegan butter

Filling

16 Pain au chocolat sitcks

8 candied orange peel sticks /
orange confit strips

Soy milk wash

Soy milk 30g

Agave syrup 15g

Pre-Check

주의 사항 P.035

Ⓗ + Ⓘ 필독!

Notice P.035

Ⓗ + Ⓘ read the suggestion!

Recipe
Timeline

한눈에 보는 레시피 타임라인

① 반죽 펴기
Final roll out

② 재단하기
Cutting

③ 성형하기
Shaping

④ 2차 발효(2hr 40min)
Final fermentation(2 hours 40 minutes)

⑤ 굽기
Baking

⑥ 식히기
Cooling

1   접힌 반죽의 양쪽을 칼로 자르고 자른 부분이 좌우로 위치하게 놓은 후 폭 28cm, 높이 34cm 크기로 밀어준다.

         TIP. 반죽의 테두리를 잘라주면 구웠을 때 결이 더 선명해진다.

2   28cm 길이의 폭은 2등분, 34cm 길이의 높이는 4등분해 14×8.5cm 크기의 사각형 8개로 자른다.

3   반죽 위에 초코 스틱과 오렌지 스틱을 1개씩 놓고 가볍게 만다.

1   Making cuts at the edges of the folded dough, place it like a book and roll it out to 28×34 cm.

         TIP. If you trim the edges of the sheeted dough, the layer is clearer when it's baked

2   Trimming any irregular edges and cut 28 cm into 2 pieces and 34cm into 4 pieces so make eight 14×8.5cm rectangles.

3   Place a piece of chocolate and orange confit at one top of each piece and it once.

4   추가로 초코 스틱을 하나 더 넣어 아래쪽으로 말아내려 성형한다.

5   성형한 반죽 윗면을 가볍게 눌러 이음매를 확실히 접착시킨다.

6   이음매가 아래로 향하게 팬닝한다.
    TIP. 발효되면서 크기가 커지므로 반죽 간격을 여유 있게 둔다.

4   Add a second stick then roll towards the bottom.

5   Press down lightly on the tops of the rolls to seal securely.

6   Place, seam side down, on a baking pan.
    TIP. They expands noticeably so leave a space between the shaped pastry.

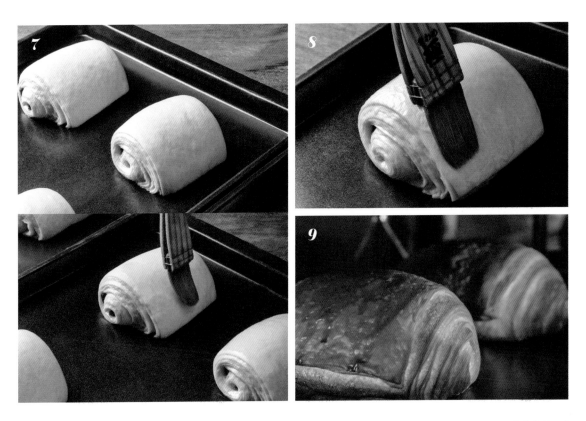

Let's Bake 2

7 반죽이 마르지 않도록 두유액을 발라주고 온도 27~28℃, 습도 75~80℃ 상태에서 2시간 40분 동안 2차 발효시킨다. 사진은 2차 발효 후 모습이다.

8 발효 완료 10분 전 오븐을 190℃로 예열을 한다. 굽기 전 두유액을 한 번 더 조심히 바른 후 오븐에 반죽을 넣고 170℃로 온도를 낮춘다.

9 13~15분간 굽고 다 구워진 빵은 완전히 식힌다.

7 Brush each rolls with the soy milk wash substitute to prevent drying out during the proof, allow the dough to rise for 2 hours 40 minutes, in a condition of 80~82°F, 75~80% to humidity.

8 When there's 10 minutes left to proof, preheat your oven to 375°F. Delicately brush the soy milk wash substitute on top of the rolls with a soft pastry brush, put the dough in the oven. Then reduce the oven's temperature to 340°F.

9 Bake for 13 to 15 minutes. Remove from the oven and let cool on the pan.

VEGAN PAIN AUX CRANBERRIES

빵 오 크랜베리

빵 오 헤장이라는 빵 들어보셨나요? 프랑스 빵집 비에누아즈리 코너에서 꼭 찾아볼 수 있는 건포도 빵으로 프랑스에서는 마니아층이 두꺼운 빵이에요. 우리나라는 건포도에 대해 호불호가 꽤 나뉘는 편이라 저는 크랜베리를 이용해 만들어봤어요. 빵 오 헤장 말고 빵 오 '크랜베리'! 묵직하지만 달콤한 빵 오 헤장에 비해 가벼운 산미가 있는 이 빵은 새콤달콤한 맛이 너무 매력적이라 먹기 시작하면 하나로는 부족할 거예요.

Have you heard of the bread called "Pain aux Raisin"? It's a raisin bread that can be found in the Viennoise section in a French bakery. This bread has quite a fanbase in France, but many people in Korea don't enjoy raisins as much. That's why I made them using cranberries! Instead of "Pain aux Raisin", "Pain aux 'Cranberry!" It is heavier but has a lighter sour taste than raisins, and I bet you that you will crave more than one after a bite of this.

Yield	8개	8 pain aux cranberries

Ingredient	**반죽 재료**	**Dough**
	비건 크루아상 반죽(P. 174) 1개	1 Laminated Croissant dough(P. 174)
	토핑	**Filling**
	건크랜베리 250g	Dried cranberry 250g
	비건 커스터드 크림(레시피 P.030)	Vegan custard cream(p. 030)
	비정제 설탕 시럽	**Simple syrup**
	비정제 설탕 100g	Unrefined sugar 100g
	물 100g	Water 100g

Pre-Check	**주의 사항 P.035**	**Notice P.035**
	Ⓔ + Ⓖ + Ⓗ + Ⓘ 필독!	Ⓔ + Ⓖ + Ⓗ + Ⓘ read the suggestion!

Recipe Timeline

① **반죽 펴기**
Final roll out

② **성형하기**
Shaping

③ **재단하기**
Cutting

④ **2차 발효(2hr 30min)**
Final fermentation(2 hours 30 minutes)

⑤ **굽기**
Baking

⑥ **식히기**
Cooling

Pre-Cook

미리 준비할 것

1. **전처리한 크랜베리** : 사용하기 하루 전에 건 크랜베리 250g에 뜨거운 물 75g을 붓고 잘 섞어 불린다.
2. **비정제 설탕 시럽** : 냄비에 물 100g과 비정제 설탕 100g을 넣고 중간 불에서 끓인다. ➪ 비정제 설탕이 완전히 녹고 끓어오르면 불을 끄고 식힌 후 통에 담아 보관한다.

1. **Cranberry maceration** : Pour 75g of hot water into the cranberry the day before using it and macerate it well.
2. **Simple syrup** : Add 100g of water and sugar each to a saucepan over medium heat. Bring it to a simmer, until the sugar has completely dissolved. Cool then store in an airtight container.

Let's Bake 1

1   접힌 반죽의 양쪽을 칼로 자르고 자른 부분이 좌우로 위치하게 놓은 후 폭 27cm, 높이 35cm 크기로 밀어준다.
       TIP. 반죽의 테두리를 잘라주면 구웠을 때 결이 더 선명해진다.

2   반죽 위에 비건 커스터드 크림을 스크래퍼로 얇게 바른다.

3   전처리한 크랜베리를 골고루 뿌려준다.

1   Making cuts at the edges of the folded dough, place it like a book and roll it out to 27×35 cm.
       TIP. If you trim the edges of the sheeted dough, the layer is clearer when it's baked.

2   Using a bent blade spatula, spread the custard cream across the surface of the rectangle.

3   Evenly scatter the cranberries over the surface of the cream.

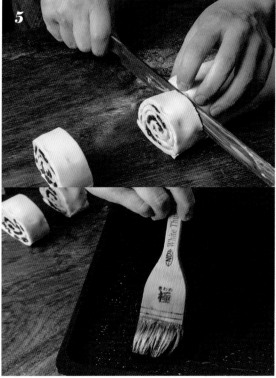

Let's Bake 2

4 위에서부터 단단하게 말아 높이 27cm의 원통형으로 성형한다.

5 기름을 바른 팬 위에 빵칼로 8등분 한 반죽을 절단면이 위로 향하도록 얹는다.
   TIP. 반죽이 말랑하다면 비닐에 감싸 냉동실에 20분 정도 냉각해 사용한다.

4 Roll the dough tightly into a 27cm long log.

5 Place the dough, divided into eight equal parts with a bread knife, on a greased pan with the cutting side facing up.
   TIP. If your dough is too soft to shape, wrap and freeze it for 20 minutes.

비건 크루아상 반죽으로 만드는 여섯 가지 빵

6   온도 27~28℃, 습도 75~80℃ 상태에서 2시간 30분 동안 2차 발효시킨다.

7   발효 완료 10분 전 오븐을 190℃로 예열한다. 발효가 끝나면 오븐에 반죽을 넣고 170℃로 온도를 낮춘다.

8   15분간 굽고 다 구워진 빵이 뜨거울 때 비정제 설탕 시럽을 바른 후 완전히 식힌다.

6   Allow the dough to rise for 2 hours 30 minutes, in a condition of 80~82°F, 75~80% to humidity.

7   When there's 10 minutes left to proof, preheat your oven to 375°F. Put the dough in the oven. Then reduce the oven's temperature to 340°F.

8   Bake for 15 minutes. Remove from the oven, brush the syrup on top during the roll is hot. Let cool on the pan completely.

VEGAN OLIVE TAPENADE & FRIED ONION PASTRIES

올리브 타프나드 & 튀긴 양파 페스츄리

몇 년 전 떠났던 스페인 여행에서 처음 올리브 타프나드를 맛보았어요. 짭조름한 것이 너무 맛있어서 뭐냐고 물어보니 올리브로 만든 거라고 하더라고요. 그 후로 집에서 자주 만들어놓고 파스타나 빵 만들 때 활용하고 있지요. 올리브의 짠맛과 튀긴 양파의 은은한 단맛이 감칠맛 나게 어우러져 참 맛있어요.

I tried olive tapenade for the first time on a trip to Spain a few years ago. I asked what this salty magic was back then, and was told that it was made of olives. After that, olive tapenade was a staple at home, whether it was eating pastas or when making bread. The salty goodness of olive tapenade and the sweet flavor of fried onions make a great combination.

Yield	10개	10 units

Ingredient	**반죽 재료**	**Dough**
	비건 크루아상 반죽(P. 174) 1개	1 Laminated croissant dough(P. 174)
	충전물	**Filling**
	양파 플레이크 1컵	1 cup onion flakes
	올리브 타프나드(레시피 P. 028)	Olive tapenade(P. 028)
	두유액	**Soy milk wash**
	두유 30g	Soy milk 30g
	아가베 시럽 15g	Agave syrup 15g

Pre-Check	**주의 사항 P.035**	**Notice P.035**
	Ⓗ + Ⓘ 필독!	Ⓗ + Ⓘ read the suggestion!

Recipe Timeline

한눈에 보는 레시피 타임라인

① 반죽 펴기
Final roll out

② 재단하기
Cutting

③ 성형하기
Shaping

④ 2차 발효(2hr)
Final fermentation(2 hours)

⑤ 올리브 타프나드 & 양파 뿌리기
Olive tapenade & onion flakes topping

⑥ 굽기
Baking

⑦ 식히기
Cooling

Let's Bake 1

1 접힌 반죽의 양쪽을 칼로 자르고 자른 부분이 좌우로 위치하게 놓은 후 폭 20cm, 높이 35cm 크기로 밀어준다.

   TIP. 반죽의 테두리를 잘라주면 구웠을 때 결이 더 선명해진다.

2 20cm 길이의 폭은 그대로 두고, 35cm 길이의 높이를 10등분해 20×3.5cm 크기의 밴드 10개로 자른다.

3 반죽의 양쪽 끝 1.5cm씩은 제외하고 중앙을 세로로 길게 자른다.

1 Making cuts at the edges of the folded dough, place it like a book and roll it out to 20×35 cm.
   TIP. If you trim the edges of the sheeted dough, the layer is clearer when it's baked.

2 Cut 35cm into 10 pieces so make 20×3.5cm 10 bands.

3 Make a lengthwise cut to within 1.5 cm of the ends.

4　반죽의 열린 부분으로 양쪽의 반죽을 넣어서 타래과처럼 꼬아 성형한다.

5　성형한 반죽은 간격을 띄워 팬닝한다.

6　반죽이 마르지 않도록 두유액을 발라주고 온도 27~28℃, 습도 75~80℃ 상태에서 2시간 동안 2차 발효시킨다. 사진은 2차 발효 후 모습이다.

4　Pass one end through the slit to twist the pastry.

5　Place it on a baking pan leaving a space between the shaped pastry.

6　Brush each pastries with the soy milk wash substitute to prevent drying out during the proof, allow the dough to rise for 2 hours, in a condition of 80~82°F, 75~80% to humidity.

Let's Bake 2

7 발효 완료 10분 전 오븐을 190℃로 예열한다. 발효가 끝난 반죽에 두유액을 바르고 짤주머니에 담아둔 올리브 타프나드를 얹는다.

8 양파 플레이크를 넉넉하게 얹은 후 물을 뿌려준다.
   TIP. 물 스프레이를 하면 구워지면서 양파 플레이크가 떨어지는 것을 막아준다.

9 오븐에 넣고 170℃로 온도를 낮춘 후 13~15분간 굽고 다 구워진 빵은 완전히 식힌다.

7 When there's 10 minutes left to proof, preheat your oven to 375°F. Delicately brush the soy milk wash substitute on top of the pastry, place the olive tapenade using a piping bag in the center of each.

8 Put onion flakes generously on the center and spray water on it.
   TIP. Spraying water is to prevent onion flakes falling off while baking.

9 Then reduce the oven's temperature to 340°F and bake for 13 to 15 minutes. Remove from the oven and let cool on the pan.

FRUIT & VEGAN CREAM PASTRY

과일 & 비건 크림 페스츄리

신선한 과일이 나오는 계절이 오면 꼭 만들고 싶은 화려한 페스츄리예요. 알록달록한 색감의 과일을 크림 위에 풍성하게 얹어 마무리하면 케이크와는 다른 매력이 있는 근사한 디저트가 되거든요. 부드러운 크림은 과일 본연의 단맛은 해치지 않고 페스츄리의 바삭함은 더 돋보이게 해줘요. 마치 시원한 바람이 부는 여름날 어느 유럽의 멋진 카페 테라스에서 휴가를 즐기는 기분이 들지요.

I always want to make colorful pastries when fresh, seasonal fruits are out. Finishing pastries by placing colorful fruits on top of the cream makes a wonderful dessert that is quite different from regular cake. The soft cream does not intrude into the sweet natural taste of the fruits, and the crunchy pastry texture helps the flavor to stand out. Whenever I try this, I always feel like I'm enjoying a vacation at a wonderful cafe terrace in Europe on a breezy, cool summer day.

Yield	8개	8 units

Ingredient	**반죽 재료**	**Dough**
	비건 크루아상 반죽(P.174) 1개	1 Laminated croissant dough(P.174)
	토핑	**Filling**
	냉장한 코코넛 크림(고형분) 300g	Hardened coconut cream 300g
	비건 커스터드 크림(레시피 P.030)	Vegan custard cream(P.030)
	신선한 과일	Fresh Fruit you want
	두유액	**Soy milk wash**
	두유 30g	Soy milk 30g
	아가베 시럽 15g	Agave syrup 15g

Pre-Check	**주의 사항 P.035**	**Notice P.035**
	Ⓗ + Ⓘ 필독!	Ⓗ + Ⓘ read the suggestion!

Recipe Timeline

한눈에 보는 레시피 타임라인

① 반죽 펴기
Final roll out

② 재단하기
Cutting

③ 성형하기
Shaping

④ 2차 발효(2hr 30min)
Final fermentation(2 hours 30 minutes)

⑤ 굽기
Baking

⑥ 식히기
Cooling

⑦ 데코레이션
Decoration

Pre-Cook

미리 준비할 것

1. **크림 토핑** : 냉장한 코코넛 크림의 고형분만 볼에 담고 휘핑한다. ⇨ 비건 커스터드 크림을 넣고 골고루 섞어 사용 전까지 냉장 보관한다.

※ **주의** : 코코넛 크림은 고형분만 사용하므로 하루 전 냉장해 크림을 충분히 굳힌다.

1. **Cream Topping** : Chilling can overnight in the refrigerator to harden and scoop the coconut cream off the top of the can into a mixing bowl and fluff up the coconut cream. ⇨ Add the vegan custard cream and mix until evenly combined, store in the refrigerator until use.

※ **Note** : Place your can of coconut milk in the refrigerator for 1 day before making whipped cream. To separate the coconut water and cream, chill the coconut milk in advance.

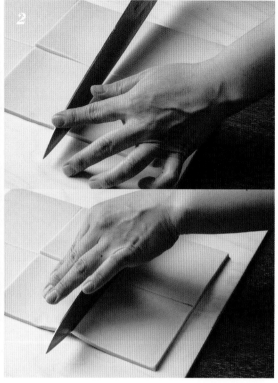

Let's Bake 1

1 접힌 반죽의 양쪽을 칼로 자르고 자른 부분이 좌우로 위치하게 놓은 후 폭 20cm, 높이 40cm 크기로 밀어준다.

TIP. 반죽의 테두리를 잘라주면 구웠을 때 결이 더 선명해진다.

2 10×10cm 크기의 정사각형 8개로 재단한다.

TIP. 반죽이 말랑하다면 비닐을 덮고 30분간 냉장 보관해 완전히 차갑고 단단해질 때 성형하는 것이 좋다.

1 Making cuts at the edges of the folded dough, place it like a book and roll it out to 20 x 40cm.

TIP. If you trim the edges of the sheeted dough, the layer is clearer when it's baked.

2 Cut into eight 10×10cm rectangles.

TIP. If your dough is too soft to shape, wrap and refrigerate for 30 minutes until it gets cold and hard.

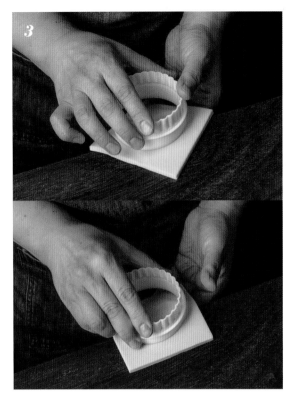
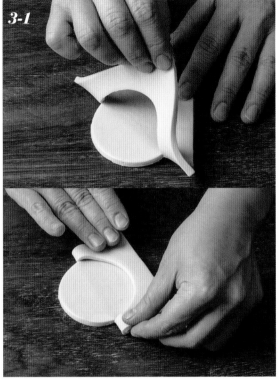

3   정사각형 반죽의 절반을 작업대 가장자리에 걸친다. 작업대 위 반죽을 쿠키 커터를 이용해 지름 6cm의 반원 모양으로 자른다.

      TIP. 기본 사각형을 원하면 성형 없이 5번 과정으로 진행한다.

3-1   반죽을 팬에 얹고 반죽 가장자리에 물을 발라준다. 잘린 쪽 가장자리 반죽을 접어 가볍게 누른다.

3   Place only half of the square dough on the edge of the workbench. Let the dough to bend naturally. Using a cookie cutter about 6cm in diameter, press the dough on the table, to cut only into a semicircle.

      TIP. If you want the basic square shape, place the dough on a baking pan without shaping.

3-1   Place the dough on a baking pan and apply water lightly to the edge of the dough with a brush. Fold the half-circle cut side across to un-cut side, to make a squared bottom and press gently.

4 정사각형 반죽을 마름모 모양으로 놓고 마주 보는 모서리에 중심선을 표시한다.

4-1 중심 부분의 1.5cm를 남기고 표시한 모서리 부분을 자른다. 같은 방향인 날개 모양의 반죽을 접고 손끝으로 중앙을 눌러 고정한다.

5 반죽 위에 비건 커스터드 크림을 얹고 온도 27~28℃, 습도 75~80℃ 상태에서 2시간 30분 동안 2차 발효시킨다.

4 Place the square dough in a diamond shape and place a ruler on the opposite corner and lightly press it to mark the centerline.

4-1 Leave about 1.5cm of the center and cut the marked line. Grab one half of each corner on same side and pull it toward the center of the square. With your fingertip, press where every point meets to fix them.

5 Put the vegan custard on the dough, allow the dough to rise for 2 hours 30 minutes, in a condition of 80~82°F, 75~80% to humidity.

Let's Bake 3

6   발효 완료 10분 전 오븐을 190℃로 예열한다. 굽기 전 두유액을 한 번 더 조심히 바른 후 오븐에 반죽을 넣고 170℃로 온도를 낮춘다.

7   13~15분간 굽고 다 구워진 빵은 완전히 식힌다.

8   식힌 페스츄리 위에 냉장 보관한 크림 토핑을 얹고 원하는 과일을 올린다.

6   When there's 10 minutes left to proof, preheat your oven to 375°F. Delicately brush the soy milk wash substitute on top of the pastries with a soft pastry brush, put the dough in the oven. Then reduce the oven's temperature to 340°F.

7   Bake for 13 to 15 minutes. Remove from the oven and let cool on the pan.

8   Fill the refrigerated coconut custard cream in the piping bag, place the cream in the center of each pastry, when cool. Place the fruits you want on top.

PART 5

VAKE VEGAN BAKING

주말 아침을 특별하게
비건 홈브런치

비건 햄버거 | 비건 샌드위치 |
비건 요거트

VEGAN BURGER

비건 햄버거

병아리콩 캔을 이용하면 쉽고 간편하게 버거 패티를 만들 수 있어요. 패티 위에 다양한 채소를 올리면 비건 햄버거가 완성됩니다. 식물성 단백질 덕분에 영양적으로 균형을 이루는 햄버거예요. 좋아하는 소스를 패티 위에 듬뿍 뿌리고 상큼한 채소를 마음껏 넣어 한입 가득 베어 물어보세요.

Use a can of chickpeas to make burger patties conveniently. Place various vegetables on top, and voilà! It's complete. Thanks to proteins inside of vegetables, this is a great nutritionally balanced food. Sprinkle your favorite sauce on the patty, and add as much fresh vegetables on top as you wish.

❶ 비건 햄버거 번 Vegan hamburger bun

Yield	10개 분량	10 buns

Ingredient	**반죽 재료**	**Dough**
	강력분 500g	Strong flour 500g
	비건 파마산 치즈 20g	Vegan parmesan cheese 20g
	소금 10g	Salt 10g
	비정제 설탕 50g	Unrefined sugar 50g
	두유 175g	Soy milk 175g
	물 175g	Water 175g
	드라이이스트 골드 7g	Dry yeast gold 7g
	비건 버터 100g	Vegan butter 100g
	토핑용	**Topping**
	참깨 & 검은깨	Sesame seeds & Black sesame seeds

Pre-Check	**주의 사항 P.035**	**Notice P.035**
	Ⓐ + Ⓑ + Ⓒ + Ⓓ +	Ⓐ + Ⓑ + Ⓒ + Ⓓ +
	Ⓗ + Ⓘ 필독!	Ⓗ + Ⓘ read the suggestion!

Let's Bake 1

1 믹싱 볼에 물을 먼저 넣고 비건 버터를 제외한 나머지 반죽 재료를 넣는다. 저속으로 3분, 중속으로 3분 믹싱한다.

2 비건 버터를 넣고 2단으로 6~7분 더 믹싱한다. 반죽 표면에 윤이 나고, 사진과 같은 얇은 글루텐이 생기면 믹싱기를 멈춘다.

3 완성된 반죽의 가장자리를 중심으로 모아서 둥근 반죽이 되게 한 후 발효통에 넣고 실온에서 60분간 1차 발효한다.

4 60분 발효 후 모습이다.

1 Pour the water and soy milk into a mixing bowl then add the remaining dough ingredients except vegan butter. Knead for 3 minutes on 1 speed then 3 minutes on 2 speed.

2 Add vegan butter and run for 6~7 minutes on medium speed. If your dough is smooth and glossy and the gluten strands have developed such like photo, stop the mixer.

3 Turn dough onto a floured surface. Fold the sides in and shape into a ball. Transfer the dough to a dough-rising bucket. Cover the bucket, and allow the dough to rise about 1 hour.

4 After 1 hour fermentation.

Let's Bake 2

5 1차 발효가 끝나면 작업대와 반죽에 밀가루를 뿌리고 플라스틱 스크래퍼를 이용해 반죽을 통에서 꺼낸다. 손바닥으로 평평하게 두드려서 일정한 두께로 만든다. 두께가 고르면 일정한 무게로 분할하기 쉽다.

6 반죽을 약 100g씩 10개로 분할하고 둥글리기 한 후 약 20분간 휴지한다.

7 휴지한 반죽에 가볍게 밀가루를 뿌리고 다시 단단하게 둥글리기 한 후 키친타월을 트레이에 얹고 축축하게 젖을 정도로 물을 붓는다. 젖은 키친타월 위로 반죽을 굴려서 반죽에 물을 소량 묻힌 후 볼에 담아 둔 토핑용 깨에 굴려 고르게 묻힌다. 이음매 부분을 아래로 가게 팬닝한다.

5 At the end of the rise, lightly flour the dough, turn the dough out onto a lightly floured surface using plastic scraper. Gently deflate the dough with your hands and make it even thickness. If you prepare the dough, it is easy to divide by the same weight.

6 Divide it into four of 250g each and shape into round. Rest the pre-shaped dough about 20 minutes.

7 After resting, lightly flour the surface of each dough and rounding them. Place some paper towels on a tray then pour the water on it to wet thoroughly. Roll the dough on the damp paper towel to moisten it and dip into a bowl of your topping before transferring to the pan. Place each bun on a baking pan seam side down.

8   2차 발효는 27~28℃ 습도 70~80%로 90분간 진행한다.

9   2차 발효 완료 10분 전 컨벡션 오븐을 190℃로 예열한다. 햄버거 번에 물을 스프레이 한 후 오븐에 반죽을 넣은 다음 170℃로 온도를 낮추어 약 12~13분 굽는다.

10   구운 햄버거 번은 식힘망에서 완전히 식혀준다.

8   Allow the dough to rise for 90 minutes, in a condition of 80~82°F, 70~80% to humidity.

9   When there's 10 minutes left to proof, preheat your oven to 375°F. Spray the water on each bun and put them in the oven. Then reduce the oven temperature to 340°F and bake for 12 to 13 minutes.

10   Remove from the oven and let cool on the rack.

❷ 비건 패티 Vegan patty

Ingredient

패티 재료

콜리플라워(냉동 콜리플라워 가능) 200g
캔에 든 병아리콩 240g(고형분)
콩물 30g
다진 파 20g
비건 파마산 치즈 20g
감자 전분 10g
강황 3g
파프리카가루 3g
후추 1g
소금 2g

패티 튀김 반죽

박력분 50g
옥수수 전분 50g
찬물 150g
소금 1g
비건 파마산 치즈 10g
후추 1g
(모든 재료를 넣어
덩어리가 없도록 잘 섞어준다)

토핑용 빵가루

비건 식빵 3~4장분
(푸드 프로세서로 갈아서 준비한다)

Patty

Cauliflower(frozen is okay) 200g
Canned chickpea, drained 240g
Chickpea can juice 30g
Chopped green onion 20g
Vegan parmesan cheese 20g
Potato starch 10g
Turmeric powder 3g
Paprika powder 3g
Ground black pepper 1g
Salt 2g

Fry batter

Cake flour 50g
Corn starch 50g
Cold water 150g
Salt 1g
Vegan parmesan cheese 10g
Ground black pepper 1g
(Put all the ingredients in a bowl and whisk
together until there are no lumps.)

Bread crumbs for Breading

Vegan plain bread 3 to 4 slices
(Grind down bread in the food processor.)

Let's Make 1

1   콩에 콩물을 넣고 푸드 프로세서로 부드럽게 갈아주고, 콜리플라워와 파는 다져서 준비한다.

2   1에 콜리플라워를 제외한 나머지 패티 재료를 섞어서 반죽한다.

3   잘게 자른 콜리플라워를 넣고 섞어준다. 푸드 프로세서로 섞을 경우 너무 곱게 다지지 않도록 주의한다.

1   In a food processor, add chickpea and chickpea can juice and process them until well mixed and smooth. Chop the cauliflower and green onion.

2   Add the rest of the patty ingredients except the cauliflower and mix well.

3   Add chopped cauliflower and mix together When mixing with a food processor, be careful not to mince it too finely.

Let's Make 2

4 120g씩 4개로 분할해서 둥글 납작한 패티로 만든다.

5 만들어둔 패티에 튀김 반죽을 입히고 비건 빵가루를 골고루 묻힌 후 프라이팬에 기름을 넉넉하게 붓고 튀기듯이 구워준다.

4 Form the mixture into four of 120g flatten ball.

5 Dip a patty in to the prepared fry batter and dip and cover everything evenly in breadcrumbs. Heat oil in a frying pan and fry the patty.

❸ 햄버거용 소스 Burger sauce

Ingredient

비건 마요네즈

두유 200g

레몬즙 또는 식초 30g

소금 6g

아가베시럽 20g

씨겨자 30g

포도씨유 200g

(바 믹서로 모든 재료를 섞는다)

치폴레 마요네즈

비건 마요네즈 200g(P.031)

다진 치폴레(훈연 파프리카) 60g

(만들어둔 마요네즈에 다진 치폴레를
고르게 섞는다)

Vegan mayonnaise

Soy milk 200g

Lemon juice or vinegar 30g

Salt 6g

Agave syrup 20g

Whole grain mustard 30g

Raisin seed oil 200g

(Blend all the ingredients with a bar mixer.)

Chipotle mayo

Vegan mayonnaise 200g(P.031)

Chopped chipotle pepper in
adobo sauce 60g

(Mix with vegan mayonnaise)

❹ 햄버거 조립하기 Assembling

Ingredient

재료

시금치 1단
양파 1개
토마토 3개
상추 약 20장
햄버거용 소스 2가지
소금
후추

Ingredient

Spinach 300g
White onion 1
Tomato 3
Lettuce 20 leaves
Burger sauces
Salt
Pepper

Let's Make

1 시금치는 깨끗이 씻어서 물기를 뺀다. 강한 불로 달군 팬에 올리브유를 두르고 시금치를 넣은 후 소금, 후추로 간해 가볍게 익힌다.

2 토마토와 양파는 슬라이스한다.

3 상추는 깨끗이 씻어 물기를 제거한다.

4 햄버거 번을 반으로 갈라서 빵 한쪽에 마요네즈를 바르고 상추, 패티, 볶은 시금치 순으로 얹고 비건 파마산 치즈를 뿌린 다음, 양파, 토마토를 쌓고, 다른 한쪽의 빵에는 치폴레 마요네즈 소스를 바르고 얹는다.

1 Rinse the spinach and drain well. Heat the oil in a skillet, add the spinach, season with salt and pepper. Cook until the spinach is just wilted.

2 Finely slice the tomato and onion.

3 Wash the lettuces and remove excess water.

4 Cut the burger bun in half and spread some mayonnaise. Then layer over some lettuce leaves, fried burger patty, spinach sautee. And sprinkle parmesan cheese on it, top the onion, tomato. Spread one side of remining bun slice with the chipotle mayonnaise and pop it on top.

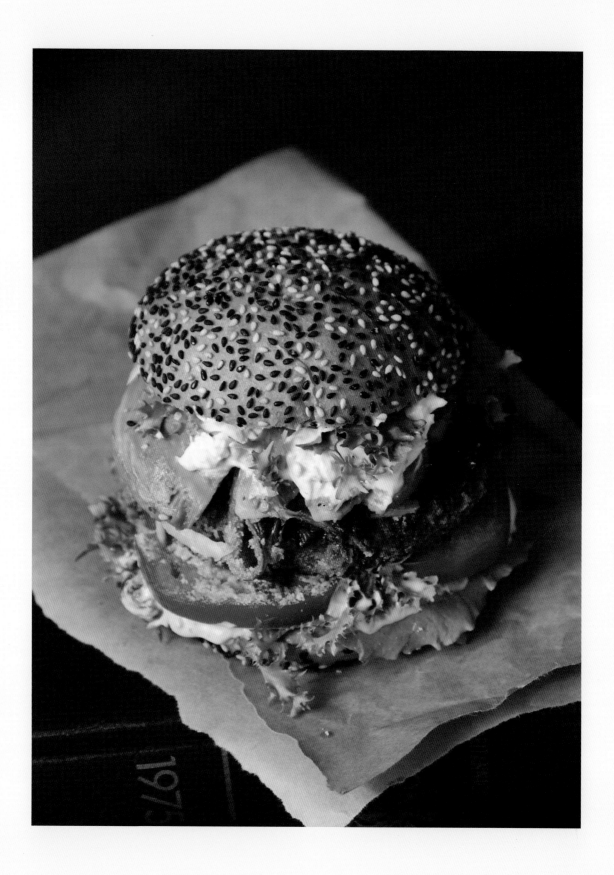

A VEGAN BRUNCH THAT COMPLETES THE WEEKEND.

VEGAN SANDWICH

비건 샌드위치

비건 치아바타 빵에 올리브 타프나드를 바르고 구운 채소를 넣어 맛있는 샌드위치를 만들어보면 어떨까요? 채소를 구우면 풍부한 단맛이 나는데, 별다른 소스 없이도 샌드위치를 맛있게 변화시켜 요. 특히 채소가 저렴한 계절이라면 양껏 사서 구운 다음 올리브유를 자작하게 부어 소금과 허브를 넣어 냉장고에 보관해두세요. 샌드위치 재료로도 만점이지만 식사 대용으로도 딱입니다. 여유 있 는 주말 오후, 돌아오는 평일을 위해 재료를 준비해두어도 좋을 것 같아요.

What happens if we spread olive tapenade on vegan ciabatta, and place some grilled vegetables on top? Yes, you guessed it right. This is a recipe for a great vegan sandwich. Grilling vegetables create a rich, wholesome sweetness that changes the taste of the sandwich without pouring any additional sauces. Buy lots of vegetables when the vegetables are ripe and inexpensive. Bake them and pour olive oil into the fridge with salt and herbs. Just keep in mind to make these in advance when you have time.

Ingredient	**재료**	**Ingredient**
	치아바타	Ciabatta
	상추	Lettuce
	토마토	Tomato
	아보카도	Avocado
	가지	Egg plant
	호박	Zucchini
	빨간 파프리카, 새싹 채소	Yellow Paprika
	올리브 페스토	Olive pesto
	비건 파마산 치즈	Vegan parmesan cheese
	비건 치폴레 마요네즈	Vegan chipotle mayonnaise

Reference	**올리브 타프나드**	**Olive tapenade**
	참고 P.028	P.028
	비건 마요네즈	**Vegan parmesan cheese**
	참고 P.031	P.031
	비건 파마산 치즈	**Vegan chipotle mayonnaise**
	참고 P.032	P.032

Let's Make

1   가지와 호박, 빨간 파프리카를 얇게 잘라서 기름 두른 팬에 굽는다. 소금, 후추로 살짝 간한다.

2   토마토와 아보카도는 슬라이스하고, 상추는 씻어서 물기를 빼서 준비한다.

3   치아바타를 반 갈라서 올리브 페스토를 발라준다.

4   상추를 여러 장 얹고 구운 가지, 구운 호박 순으로 얹은 다음 비건 파마산 치즈를 뿌린다. 그 위에 구운 파프리카, 토마토 슬라이스, 아보카도를 올려준다.

5   빵의 다른 한 면에는 치폴레 마요네즈를 바르고 얹는다.

1   Finely slice the eggplant and zucchini, deseed and slice up the paprika. Heat a little oil in frying pan and fry the vegetables then season with sea salt and black pepper.

2   Finely slice the tomato and avocado, wash the lettuces and remove excess water.

3   Cut the ciabatta in half and spread on some olive tapenade.

4   Then layer over some lettuce leaves, fried vegetables, sprinkle parmesan cheese on top. Arrange the tomato, avocado on it.

5   Spread one side of remaining ciabatta slice with the chipotle mayonnaise and pop it on top.

A VEGAN BRUNCH THAT COMPLETES THE WEEKEND.

VEGAN YOGURT

비건 요거트

요즘은 대형 마트나 온라인 구매처에서 비건 요거트를 어렵지 않게 찾을 수 있어요. 시중에 판매하는 비건 요거트 제품에 두유를 섞어 먹어도 좋고, 비건 요거트 스타터를 활용해 집에서 간편하게 만들 수도 있어요. 아가베 시럽이나 과일, 시리얼 등을 곁들여 가볍게 아침을 즐기세요.

You can easily find vegan yogurt at large supermarkets or online these days. You can mix soy milk with vegan yogurt products that you buy, or you can make it yourself with vegan yogurt starters. Enjoy a light breakfast with agave syrup, fruits, cereals and more.

Ingredient

재료

두유 950㎖ 1팩
비건 요거트 스타터 1포(약 2g)

요거트 메이커

Ingredient

Soy milk 950㎖(room temperature)
A packet of vegan yogurt starter culture
(about 2g)
Yogurt maker

Let's Make

1. 실온에 놓아둔 두유에 요거트 스타터를 넣고 잘 섞어준다.

2. 요거트 메이커에 뜨거운 물을 넣고 두유액을 넣은 뒤 10~12시간 따뜻한 곳에 두어 발효시킨다. 계절에 따라 실내 온도가 다르므로 발효 시간이 달라질 수 있다.

3. 통을 기울였을 때 요거트가 흘러내리지 않으면 완성된 것이다. 발효된 요거트는 냉장고에 넣어 4~5시간 보관 후 먹는다.

   TIP. 비건 스타터가 없으면 시중의 비건 요거트를 스타터로 사용해도 좋다. 그럴 경우 요거트 스타터 1포 대신 요거트 1통(약 100g)을 넣는다. 요거트 메이커가 없는 경우 전기밥솥의 보온 기능을 활용해도 된다.

1. Pour lukewarm soymilk with yogurt starter culture in a container and stir well.

2. Pour the hot water in your yogurt maker and place the container in it and let it incubate for 10 to 12 hours. Since the indoor temperature varies depending on the season, the fermentation time may also vary.

3. If the yogurt doesn't run down when the container is leaned, it's done. Refrigerate it about 4 to 5 hours before serving.

   TIP. You can also use a yogurt as your starter. In that case, use one pot(about 100g) of yogurt instead of a packet of yogurt starter culture. You can use instant pot or rice cooker to make your yogurt.

VAKE VEGAN BAKING